MANAGEMENT OF TOXIC SUBSTANCES IN OUR ECOSYSTEMS
Taming the Medusa

MANAGEMENT OF TOXIC SUBSTANCES IN OUR ECOSYSTEMS
Taming the Medusa

Edited by
Barney W. Cornaby

ANN ARBOR SCIENCE
PUBLISHERS INC / THE BUTTERWORTH GROUP

230 Collingwood, P.O. Box 1425, Ann Arbor, Michigan 48106

Library of Congress Card Catalog Number 81-67257
ISBN 0-250-40412-5

Manufactured in the United States of America

Butterworths, Ltd., Borough Green, Sevenoaks, Kent TN15 8PH, England

Preface

The symposium "Taming of the Medusa: Toxic Substances in Our Ecosystems" was held in Columbus, Ohio, March 7–8, 1980. It was one in a series of ecology symposia under the auspices of the Ohio Academy of Science (Ecology Section), Battelle Columbus Laboratories and other institutions to explore important trends in ecological and environmental sciences. Each symposium has provided a forum of well-informed nationally and internationally recognized experts and has resulted in an enlightened thumbnail sketch of a given topic. Since toxic substances had come of age, so to speak, plans in early 1979 called for a symposium on the ecology of toxic substances.

The ecology symposia, with a legacy of interesting and useful topics, are usually held in late winter at Battelle's auditorium. Previous themes have been: Ecological Succession (1973), Biological Implications of Strip Mining (1974), Energetics and Fitness (1974), Environmental Impact Assessment: the Role of Biologists (1976), Stress Effects on Natural Ecosystems (1977), and Training and Personnel Trends in Ecology and Environmental Sciences (1978). Because we wanted a larger readership for this vital topic, the outlet for the 1980 proceedings is different. The new policy of the Academy permitted me to work with Ann Arbor Science Publishers in preparing this volume and bringing this timely symposium proceedings to the attention of biologists, engineers and other interested persons in government, industry and academia.

The sponsors of the 1980 symposium deserve mention. In alphabetical order they are: Battelle Columbus Laboratories (Bioenvironmental Sciences Section), The Ohio Academy of Science (Ecology Section), Kent State University, Miami University (Institute of Environmental Sciences), Oberlin College, The Ohio Biological Survey, The Ohio State University (Environmental Biology Program and Department of Zoology) and The University of Akron. I was assisted by a program committee: Dr. G. Dennis Cooke (Biology, Kent State University), Dr. Barbara A. Schaal (Botany, The Ohio State University) and Mr. Lynn E. Elfner (executive officer, Ohio Academy of Science). Drs. Kenneth M. Duke, Gerald L. Fisher and Anna D. Barker of Battelle arranged financial support for my role in both the March 1980 symposium and in editing all the manuscripts and preparing the book. Dr. Kenneth M. Duke critically reviewed all of the manuscripts.

This volume presents the opening remarks and the six papers given at the symposium. The third paper, by O'Neill and Waide, was not presented orally because Dr. O'Neill became ill on the day preceding the meeting and could not travel. The papers generated a great deal of interest, with lively question-and-answer periods following each three presentations. Although this dialog is not printed, it is available on tape from the Ohio Academy of Science. I prepared the concluding remarks in March 1981 from notes made at the symposium, after I had edited all of the manuscripts and considered their impact.

Barney W. Cornaby

Barney W. Cornaby is a Senior Research Scientist with Battelle Columbus Laboratories. He received his PhD in ecology/entomology from the University of Georgia, his MS in zoology/statistics and his BS in zoology/Portuguese from Brigham Young University. He has also studied at the University of Utah and Universidad de Costa Rica.

Dr. Cornaby's research interests at Battelle have been concentrated on technical evaluation of health/ecological effects from dams, pipelines, highways, chemical plants and power plants. Previously, he conducted environmental research in all major ecosystem types in the United States, including wetlands, estuaries, deserts, grasslands, alpine systems, forests and farmlands. He recently directed a large project involved with toxicological and ecological research in Venezuela. His current research is focused on the development and use of nonhuman organisms in early warning systems to protect humans at waste disposal and other sites, and on the design and evaluation of new products, especially those involving toxic substances.

Dr. Cornaby is a member of the American Association for the Advancement of Science, Ecological Society of America, Entomological Society of America and the Tropical Biology Association. He has published 18 articles in professional journals and numerous technical reports, and has presented more than 30 papers at national and international meetings.

Contents

1. Opening Remarks 1
B. W. Cornaby

2. Toxic Substances: Clear Science, Foggy Politics 5
G. M. Woodwell

3. Environmental Carcinogens: The Human Perspective . . 19
N. A. Reiches

4. Ecosystem Theory and the Unexpected: Implications for Environmental Toxicology 43
R. V. O'Neill and J. B. Waide

5. Paradigms in Multiple Toxicity 75
P. D. Anderson

6. Development of New Bioassay Protocols 101
K. M. Duke and R. G. Merrill, Jr.

7. Toward Improved Control of Toxic Substances 121
G. D. Rawlings

8. Closing Remarks 145
B. W. Cornaby

Author Index 151

Subject Index 157

1

Opening Remarks

Barney W. Cornaby
Health and Environmental Sciences Section
Battelle Columbus Laboratories
Columbus, Ohio

THE MEDUSA MYTH

The three terrible sisters from Greek mythology, the Gorgons, one of whom was named Medusa, were winged creatures, having the form of young women with glaring eyes, tusks for teeth and serpents for hair. Theirs was a petrifying power, because their appearance was so hideous that whoever looked at them was turned to stone.

Medusa was mortal, while her sisters, Stheno and Euryale, were immortal. The myth is thousands of years old, and according to some writers the sisters lived on the Atlantic side of Africa. At their living place, one could imagine rain-worn shapes of men and wild beasts whom they had petrified. Clearly, it would require astuteness and perhaps even the help of deity to control them.

Perseus, son of Zeus, was sent to fetch Medusa's head. Perseus was aided in his quest by various mythical persons who gave him winged sandals, a large sack and a cap that made him invisible. He also received a special curved sword to use for his gruesome task and a highly polished shield to protect him from the sisters' gaze.

The Gorgons were asleep when Perseus arrived at their retreat. Looking at them through the polished shield, he decapitated Medusa. He put her head in the sack and fled with the other two Gorgons in pursuit. However, they failed to catch him because he was made invisible by the magic cap. While flying over present day Libya, blood from the head seeped through the sack and dripped onto the desert floor, where the heat spawned many snakes.

Later, the head of Medusa, which retained its petrifying power, was placed on warriors' shields and elsewhere as a protective talisman. Athena's shield, for example, showed the Medusa head.

Paintings and sculptures depict Medusa with a variable number of and range of conditions for the snakes. In some cases the snakes were many and well-fed; in other cases the snakes were few and skinny. Regardless, there were always sufficient snakes to provide a frightening experience to the viewer.

TOXIC SUBSTANCES AND THEIR PROBLEMS

Our major concerns with toxic substances are as numerous as the snakes on Medusa's head. There are about four million registered chemical compounds, and more than 30,000 of them are used in commerce. However, the kinds and quantities of potentially toxic chemicals appearing on the market are increasing each year. In 1976 about 243,000 chemical substances were recorded in the Inventory Candidate List for the U.S. Toxic Substance Control Act, and hundreds of substances have been added since then. However, the 1977 edition of the Registry of Toxic Effects of Chemical Substances contains data for only 26,000 entries, pertaining mainly to humans and laboratory animals. Clearly, there is a growing backlog of chemicals whose toxic potential is not known.

Likewise, the volume of chemicals produced is increasing. Figures from the U.S. International Trade Commission show that annual production of synthetic organic chemicals in the United States increased 50 to 184 billion pounds from 1959 to 1974. Production of other chemical products has similarly increased. For instance, pesticide production in the United States has more than doubled—from 0.6 to 1.4 billion pounds—annually during the same period. These trends continue.

One of our civilization's major challenges is to manage properly this rapid proliferation of new substances. True, these new substances provide many benefits:

- pesticides to enhance agricultural production
- synthetics to replace more expensive raw materials
- pharmaceuticals to extend life
- fuels to provide more energy.

The list of contributions is long. However, these substances also provide new sources of potentially hazardous effects: toxicological problems are surfacing with alarming frequency. Almost daily, we learn disturbing accounts:

- high concentrations of heavy metals being found in fish
- small children being exposed to lead-containing paints
- workers in a Kepone chemical plant and nearby residents displaying symptoms of poisoning
- the chemical TRIS (used in flame-resistance pajamas) as a possible link to cancer
- the perils of improper chemical waste disposal rising to haunt us at tens-of-thousands of locations worldwide.

The list of problems is long.

THE ANALOGY

How can we battle this contemporary Medusa? These proceedings provide some of the answers. First, the enormity of the problem will be sketched by George M. Woodwell. He describes, as it were, the head of the Medusa and gives the charge for battle. Then, one by one each successive speaker symbolically severs the head of a snake. The first two speakers discuss different but interrelated perspectives on the toxic substance problem: human populations and ecosystems. They deal with the snakes of epidemiologic studies and extrapolation from animal toxicity and chemical data to the intervention of human cancer patterns (Nancy A. Reiches) and the hierarchical and biogeochemical cycles approach to measuring toxic effect to ecosystems at the system, not the component, level (Robert V. O'Neill and Jack B. Waide). With this preparation, three more snakes are attacked. They follow: synergistic and antagonistic effects of chemical mixtures (Perry D. Anderson), protocols of biological tests for documenting effects of emissions (Kenneth M. Duke and Raymond G. Merrill, Jr.) and management strategies for controlling certain substances (Gary D. Rawlings). Together these papers provide a logical series that explain how toxic substances are being or can be managed in our ecosystems. Finally, I return to provide a synthetic view of all the papers and to evaluate how successful we were in taming the Medusa in the Closing Remarks.

2

Toxic Substances: Clear Science, Foggy Politics

George M. Woodwell
The Ecosystems Center
Marine Biological Laboratory
Woods Hole, Massachusetts

In 1972 Humpstone examined contemporary pollution problems through the eyes of a lawyer [1]. He observed that in the middle eighteenth century, more than 200 years ago, another lawyer, Blackstone, had drawn on an ancient Christian maxim as the foundation of the law of nuisance: *sic utere tuo ut alienum non laedas* (use your own property in such a way as not to injure another's). Blackstone offered examples: "To build a house so close to another's that rainwater from your roof spills onto his, is to commit a nuisance." So is keeping hogs so close to another's house "that the stench of them incommodes him and makes the air unwholesome." One responsible for a nuisance was "to find some other place to do that act where it will be less offensive."

In 1980 we live with 4.4 billion other humans, a complex technology and almost infinite aspirations on a small, green planet. Our 4.4 billion people will become 6.0 billion in less than 20 years, barring nuclear war or other catastrophe. Those now living who have a 50-year life expectancy can look forward to living in a world with 9.0 billion people, if current trends are not changed. Places to hide are already few; what was remote and therefore safe in the past is no longer safe. There is no more powerful example of the limits we are now encountering than that offered by toxic substances.

The 1979 Annual Report of the U.S. Council on Environmental Quality [2] reports that more than 4 million chemicals have been registered with the American Chemical Society since 1965; that more than 43,000 chemicals, not including pesticides or drugs, are listed by the U.S. Environmental Protection Agency (EPA) as subject to the Toxic Substances Control Act (TSCA); and that there is a continuing rise in deaths from cancer of 0.5% annually, with the number of new cases of cancer rising at 1.6%/yr (80–90% of these are attributed to environmental factors). Industrial activities that produce toxins at such a scale are an obvious threat to life itself. How are such materials to be managed? What experience can we invoke in their management?

NATURE OF THE PROBLEM

Examples of the problems with toxins are overwhelming. One of the most spectacular in recent years has been the Love Canal incident in Niagara Falls, NY, where a chemical company dumped and buried diverse toxic wastes over a period of years. Houses were built on the land, and in a time of heavy rains, toxins moved with the groundwater into cellars and ultimately seeped to the surface to foul air and water and to threaten human health. There may be several thousand such dumps in the United States and several hundred that could offer similar problems. Love Canal alone had cost the taxpayers more than $27 million by the early months of 1980: money spent to move the families who bought houses on this land, to control the toxic substances and to make limited restitution. The problem continues, and many more families seem certain to be affected. Furthermore, the money spent to date has done nothing to remedy the biotic effects. To emphasize the magnitude of the problem, the federal government attempts to control some 43,000 different pesticides produced by 7400 manufacturers, whose output is an estimated 1 billion pounds of toxins with a total value of $2 billion [2]. The management of sewage and industrial wastes is not included in these statistics.

The problems in controlling these substances embody nearly all the elements of the contemporary crisis of environment: growth, profits, and economic and political power. The issues reach further to challenge what many consider as reason itself, even to challenge compromise, the very basis of politics in the democracies. Small wonder that the topic is contentious.

The root of much of the uncertainty among those who attempt to manage toxins lies in the fundamental assumptions on which management is usually attempted. A common approach is to emphasize that toxicity is a matter of concentration or quantity, not quality, or an intrinsic characteristic of the substance. Pollution, too, is considered a question of degree. Implicit in the analysis is a system of thresholds below which tolerance exists, sometimes formalized in specific instances as "assimilative capacity." This concept is applied to organisms and to nature as a whole. Although convenient, attractive and apparently reasonable, I suggest that the concept is misleading, if not simply wrong, and that it is especially misleading when used in nature. I argue, moreover, that the present system for managing toxins is inadequate to protect man from toxication and the biota from impoverishment.

The key point should be sufficiently well known to be trite: the world is a biotic system, the product of a biotic evolution that has produced 3–10 million different kinds of plants and animals, each very precisely attuned for survival to a narrow set of physical, chemical and biotic conditions. Despite man's success in turning nature to his own purposes, we still live as guests in a biosphere dominated by natural communities that operate according to a complex and poorly recognized set of laws that have their basis in evolutionary processes. We are learning now that the organisms in these communities may respond to one another through chemical signals that are effective even when concentrations are as low as one part per trillion. In addition, we know that evolution breeds diversity in form and function: virtually every circumstance and every resource that can be exploited to sustain life has been exploited in the fullness of four billion years of constant evolutionary testing. Sudden changes in environment, whether they are physical, chemical or biotic, will bring sudden changes in the biotic systems of that place. Sudden change is as disruptive and as expensive in natural systems as it is in a watch, a factory or a transportation network.

TWO EXAMPLES

The best examples are from experience with two groups of toxic substances: ionizing radiation and pesticides. Ionizing radiation is an example of an inadvertent waste product of technology that happens to have powerful biotic effects. Pesticides, in contrast, are produced and used because of their biotic effects.

Ionizing Radiation

Our experience with radiation as an important worldwide contaminant began with the discharge of the first fission bombs in 1945 [3]. The scale of the problem did not become clear until almost ten years later, when, in the spring and summer of 1954, the United States completed a series of tests of bombs at Bikini Atoll in the western Pacific. The first of these tests, BRAVO, was especially notable because the fallout moved eastward instead of westward and reached Rongelap Atoll, where 65 Rongelapese received about 50% of a mean lethal exposure before they could be evacuated. In all, the series of tests that spring contaminated 10,000 mi^2 of the Pacific with fallout that would have been lethal to man had he been there. For many weeks, fish landed in Japan could not be sold because they were sufficiently radioactive from the oceanic contamination to be judged inedible. Oceanographic surveys by Japan and the United States confirmed the extent of the contamination and the fact that radionuclides were being accumulated by the fish. Continued testing of bombs in the atmosphere introduced sufficient additional radioactivity into the troposphere that radioactive rains were detected in various parts of North America.

The tests, whatever one may think of them in hindsight, offered a remarkable series of tracers for study of atmospheric and oceanic circulation. Elaborate research programs started by the Atomic Energy Commission over a period of more than 10 years provided a wealth of information about the circulation of the atmosphere, including the transport of particles and their deposition on the earth's surface. The evidence showed, for example, that small particles introduced into the troposphere in the middle latitudes are carried around the world in two to three weeks. Particles are removed from the atmosphere by precipitation, with the amount of removal diminishing as the rainfall continues. Even small particles, (e.g., pollen grains), that can be windborne, travel in these atmospheric patterns. Deposits of radioactive debris were greatest in the middle latitudes because precipitation is high there. Areas that lie in common storm tracks, such as New England, receive especially large deposits of materials that are carried in the atmosphere. Finally, the transfer of particles in the atmosphere between the Northern and Southern hemispheres is limited; the deposits of radioactivity in the Southern Hemisphere were approximately 10% those in the Northern Hemisphere for many years. Apparently, latitudinal exchanges of air and particulate matter between hemispheres are very much slower than longitudinal transport within the hemispheres.

There is ample reason to believe that any substance that can be vaporized or distributed as small airborne particles can be expected to have a regional, possibly a worldwide, distribution. Thus, there is no mystery as to why traces of pesticides or polychlorinated biphenyls (PCB) are found in mountain lakes and in glacial ice many miles, sometimes thousands of miles, from where they could have been used.

But even more important, we learned that radioactive elements, such as strontium-90, cesium-137 and iodine-131, that are similar to elements essential for life can be accumulated from very low concentrations in nature to high concentrations in living systems. The lesson should not have been necessary because we know that living systems exist by processes that result in the accumulation of elements necessary for survival. That, too, is an evolutionary legacy.

All of this evidence emphasizes that toxins that have a residence time in the atmosphere of hours to days can be carried worldwide in days to weeks by atmospheric transport alone, and that biotic mechanisms, in addition to physical mechanisms, act to return toxins to places where they again affect man directly. This experience alone should have been enough to destroy the arguments about dilution, including the use of the sea for disposal of persistent toxins. It was not.

There were further lessons from radioactivity. They lie in the realm of effects. The biotic hazard from ionizing radiation is in the ionization of biologically important molecules. The most important biotic molecules are those that carry the information for operating the organism, the genes. The production of mutations in man is commonly thought to be deleterious because we consider each individual important and are unwilling to sacrifice lives in favor of genetic improvement of the group as a whole. Mutant genes, most of which are deleterious, are not systematically eliminated from the population by selective pressures operating against individuals, at least not in places served by modern medicine. We make extraordinary efforts to avoid the loss of individuals, and the mutants may, therefore, accumulate. Man is unique in this respect among other species of the earth. All other species are subject to elimination of mutants through early death of the individuals that carry them. The objective in managing the hazards of radioactivity is to protect man, not from somatic effects, but from an increase in the rate of mutation in his germ cells. If we are successful in protecting man, all other species will have been protected. The emphasis on human safety is appropriate and provides adequate insulation for the rest of the biota. This relationship is special and is not shared by other toxins that have different mechanisms of biotic effects. Thus, ionizing radiation is not an appropriate model for management of these other toxins.

Pesticides

The second series of studies to which I call attention is the now almost monstrous literature on pesticides. The best consolidated body of that literature treats the persistent pesticides, especially dichlorodiphenyltrichloroethane (DDT).

The insecticidal properties of DDT were discovered in 1939 by Muller in Basel, Switzerland. The DDT molecule is stable under a wide range of conditions and retains its insecticidal properties for days or weeks in the field. In addition, its toxicity to mammals is low. It was used extensively and effectively during World War II to combat lice and mosquitoes. After the war, it was used widely in agriculture and public health programs.

The very characteristics that made DDT an effective insecticide (its broad toxicity to arthropods and persistence) contributed to the series of problems that led to the banning of its use in the United States and many other nations. Despite a clear warning published in 1946 that DDT should not be used in marshes or other places where it could move into waterways [4], it was by 1966 widely used for control of the salt-marsh mosquito in many areas around the world. DDT was being used in aerial spray programs to attempt to control such widespread pests at the gypsy moth in eastern North America and the spruce budworm of the boreal forest. Residues could be found in virtually every organism on earth, terrestrial and marine, including Antarctic crabeater seals and penguins [5]. Most people were carrying residues of a few to several tens of parts per million in their fat.

Various studies of the distribution of residues of DDT and its decay products showed that although concentrations in the general environment might be extremely low (less than one ppb in water), concentrations in the biota ranged up to levels that had been shown to be lethal under certain circumstances. Concentration factors exceeding 10^5 were common. The effects were especially conspicuous in populations of carnivorous birds where reproductive failures were widespread. Parallel effects occurred in amphibians and fish. Measurement of such changes is always difficult, and it is probable that only a fraction of the effects caused by this general contamination with DDT were actually documented.

There were several lessons from the DDT experience. First, persistence in a pesticide is a double-edged sword. While it is desirable for a pesticide to retain its effectiveness, the longer it does so, the greater the probability that its residues will migrate and accumulate in places where it presents a hazard to man or other organisms. Second, there is no possibility of using a persistent pesticide in agriculture without contaminating the human food chain. A farmer who sprays DDT on potatoes is simultaneously contaminating the pasture of the farmer nearby, and the residues will appear in the milk of the neighbor's cow. Third, when any pesticide is released into the atmosphere, its wide distribution as small particles or vapor in the lower troposphere must be expected as explained in the section on radionuclides. Indeed, DDT residues were carried around the world. While the world circulation has never been defined in sufficient detail, there is abundant evidence of aerial transport of DDT residues and their accumulation in biotic systems remote from any place where DDT was used. Finally, the use of broad-spectrum pesticides that affect many different kinds of organisms can be expected to result over time in the evolution of resistant strains of the pests and the progressive reduction in the effectiveness of the insecticide. This transition has been observed repeatedly for DDT and most other pesticides.

While most contemporary pesticides are not as persistent as DDT, and many do not share the sharp differential solubility of DDT between fats and water, the DDT example remains the classic case. The persistence and toxicity of DDT residues were such that they migrated out of agricultural areas into adjacent terrestrial and aquatic systems and affected them. The net effect was a general toxication of natural ecosystems and impoverishment in some cases. The increments of change due to such toxication are small and very difficult to prove. In the case of DDT, there was overwhelming evidence, although some of the proponents of DDT use argued that the evidence was far from overwhelming. Nonetheless, after a protracted series of court cases and hearings extending through 1972, the use of DDT within the United States was prohibited by Ruckelshaus, then Administrator of the EPA.

The most important reason for banning DDT was the fact that its residues were accumulating in the biota with deleterious effects. The direct threat to human health was clearly secondary. Detailed study had shown that there was substantially no means to regulate DDT use that would prevent its accumulation in soils and biotic systems to the point where it affected bird and fish populations. Through these routes, DDT was known to accumulate in humans; although this could not be proven to be clearly deleterious, it was certainly of no advantage to humans.

The DDT experience has profound implications. The government itself found that, contrary to most experience in government, there was no intermediate solution. There was no way of solving the problem without eliminating DDT. It is still awkward to say that there is no room in the world for a substance that has proven as useful as DDT, but there was then, and is now, no basis for establishing a safe level of usage. What is the threshold for effects when residues can be accumulated from water to concentrations that may be several hundred thousand times greater than those in the water? What is the assimilative capacity of any environment for DDT residues? The only reasonable action was elimination of DDT. But we seem not to have learned from the experience.

Examples of the failures of current policy management of other persistent toxins are accumulating daily. A recent front-page article in the *New York Times* announced that as many as 10,000 wells on eastern Long Island may have been contaminated through use of pesticides for culture of potatoes. The contamination is irreversible in any time-frame of interest to us. Virtually every bluefish caught on the eastern seaboard of the United States contains residues of PCB that have been released into estuarine areas or areas where they can be carried by the atmosphere to the oceans. These and other residues obtained directly through food contribute to the burden of chlorinated hydrocarbons that is now even a part of human milk. If this same burden appeared in milk produced for sale, there would be a regulatory requirement that the milk be removed from interstate commerce.

WHAT WE MUST DO

The circumstance at the moment is that, despite our experience, we have developed a system for managing toxins, including industrial wastes and pesticides, that is (1) permissive in the extreme; (2) based on thresholds and assimilative capacities; and (3) now leading to a general contamination of human beings. Modification of this management system becomes more difficult daily because the financial considerations, already great, continue to grow. Industrialized agriculture is firmly committed to use of pesticides of various types. Other segments of industry find the most serious financial problems associated with abridgment of their license to allow wastes to diffuse into the general environment, casting off those costs to be shared by the public at large while the profits are sharply focused within the industry.

This circumstance—the protection of the common interest in shared resources—is precisely the circumstance that governments must be expected to address. There is no other agency, no financial gradient, no force of self-regulation that can be expected to be effective. While we may deplore additional interference by government in the economic system, there is no alternative.

The lessons seem clear. First, biotic systems have determined and are now dependent on the chemical composition of the environment. Second, the chemistry of the environment is vulnerable to change from human activities, and the changes are increasingly global as opposed to local. Third, virtually any substance that has a low vapor pressure or is released as small particles into the lower atmosphere is subject to worldwide dispersion in days to weeks. Fourth, substances that are absorbed into the biota may accumulate to concentrations that are hundreds to thousands of times higher than concentrations in air or water. This capacity of biotic systems to accumulate biotically active substances offers strong evidence against the assumption that toxic substances can be managed on the basis of thresholds and assimilative capacities. Small changes in the chemistry of the environment produce changes in the biota—changes that may be very large in proportion to the magnitude of the chemical change.

These conclusions have been drawn from our extensive experience with radioactivity and pesticides. They apply widely to toxins, although one may question whether they can be applied with equal validity to elements used in life. For nutrients, however, additional evidence is abundant that the biota is sensitive to shifting ratios: small changes in the availability of phosphorous or nitrogen bring large changes in the composition of natural communities on both land and water. Again, the principle seems to be that there is a continuum between the biota and chemistry of the environment. The threshold concept has no role here.

The evidence for virtually everything I have said here about toxins and nutrients alike is partial. It is based in part on observable fact and in part on accumulated experience in study and interpretation of natural communities. If the evidence is as important as it seems, the current system for management of the chemistry of the environment is certain to lead to general toxication, impoverishment of nature and progressive contamination of human food webs. The increasing number of accidents, such as the Love Canal disaster, should in itself be adequate evidence of the severity of the problem.

Solutions are always more difficult than recognition of the

problem. How can we continue to have the advantages of the contemporary industrially based societies and simultaneously protect the earth's biotic systems from progressive impoverishment through toxication?

The answer is that there is no alternative to the development of closed systems for management of toxins. To do this may require development of a new body of knowledge based on original research. The transition is already underway in industries forced for economic reasons to recover wastes and for environmental reasons to eliminate toxic releases. The European community is supporting conferences and research on low-level waste technology. A small further step would be toward support of research on how to increase the degree of closure of existing industrial, domestic, urban and agricultural systems. The transition, sometimes called "reindustrialization," has many elements, none beyond reason or the realms of possibility. Pesticides can be developed that are short-lived and affect only the pest. Self-contained domestic water systems are possible. They would eliminate the need for elaborate sewage collection systems that release wastes into water bodies, and would simultaneously reduce or eliminate the need for large supplies of domestic water. The control or elimination of industrial wastes is a proper cost attributable to the product, not to be borne by the public at large.

The steps seem simple, obvious, and only fair—a natural consequence of the application of the ancient Christian doctrine, cited by Blackstone as the foundation of the law of nuisance; *sic utere tuo ut alienun non laedas.*

REFERENCES

1. Humpstone, C. C. "Pollution: Precedent and Prospect," *Foreign Affairs* 50:325–358 (1972).
2. Council on Environmental Quality. "Environmental Quality—1979; Tenth Annual Report of the Council on Environmental Quality," U.S. Government Printing Office (1979).
3. Hines, N. O. *Proving Ground: An Account of the Radiobiological*

Studies in the Pacific, 1946–1961 (Seattle, WA: University of Washington Press, 1962).

4. Cottam, C., and H. Higgins. "DDT and Its Effect on Fish and Wildlife," *J. Econ. Entomol.* 39:44–52 (1946).
5. Sladen, W. J. L., G. M. Menzie and W. L. Reichel. "DDT Residues in Adelie Penguins and a Crabeater Seal from Antarctica," *Nature* 210:670–673 (1966).

3

Environmental Carcinogens: The Human Perspective

Nancy A. Reiches
Comprehensive Cancer Center and Department of Surgery
The Ohio State University
Columbus, Ohio

The modern era in medical research has witnessed outstanding advances in the understanding and control of many infectious diseases. Diseases such as tuberculosis and smallpox that at one time claimed large numbers of lives, especially in young age groups, are now infrequently encountered. However, the resulting increase in life expectancy and the growing complexity of the industrial-chemical environment have presented us with a new set of human health problems, questions and risks. Today, much of our emphasis has shifted from evaluations of exposures that almost always result in untoward outcomes to the examination of the modest risks posed by long-term, low-level exposures to chemicals and other hazards associated with our technologies.

This situation has given rise to one of the most challenging problems in contemporary epidemiology and preventive medicine, namely, the observation that most cancers appear to be induced by environmental factors, rather than by purely genetic factors. To the extent that carcinogenic agents and processes originate outside the host, there is potential for intervening in the environmental network in much the same way as we broke the web of causality for infectious diseases.

The cumulative risks from the industrial-chemical environment, both additive and synergistic, are believed to be major contributors to the alarming increase in human cancer incidence and mortality. Currently, cancer is the second leading cause of death, accounting for approximately 20% of all deaths. It is estimated that more than 800,000 new cases of cancer will be diagnosed in the United States in 1981, with over 40% of these cases accounted for by carcinomas of the lung, colon and female breast. Under the age of 55 years, there are more deaths from cancer than from heart disease. Furthermore, the incidence of cancer increased at a rate of 1.3%/yr for white males and 2.0%/yr for white females between 1969 and 1976 [1]. Current hypotheses suggest that these increases are attributable to changes in environmental factors. As the rates of malignant disease have risen in nearly all population subgroups, it has become tenable to believe that even risks posed by low levels of environmental contamination may be of singular importance to human health.

This chapter focuses on the protection of humans from undue exposures to carcinogenic substances, especially low-level and chronic exposures. First, two major approaches to evaluating human risk—animal studies and direct observations on human populations—are presented, and methodologic problems in establishing cause-effect relationships between toxicants and disease states are briefly described. Second, the issues in the study of human diseases from chronic exposure to organic chemical constituents of drinking water are treated both in their historic and contemporary perspectives. The approach is advanced as a classic epidemiologic model, demonstrating both the difficulties and the successes of human observational analysis in disease control. Finally, the need for convergence of chemical, toxicologic and epidemiologic evidence is explained. The need for balance is emphasized between the probabilistic approach of science and the deterministic approach of regulatory agencies.

Eventually, scientific evidence and its social implications should result in prudent policy for intervening in the natural history of human noninfectious diseases, especially those originating from the chemical environment.

APPROACHES TO THE EVALUATION OF CARCINOGENS

A very small number of substances are confirmed human carcinogens. Some are substances used in industry that have resulted in high occupational exposures, e.g., arsenic, cadmium, chromium and vinyl chloride. Others are substances used by individual choice, such as alcohol and tobacco. Still others are substances in the ambient environment. However, even though the list of known human carcinogens is somewhat short, an increasing number of chemical compounds has been determined to be animal carcinogens. When a substance is carcinogenic in animals, it suggests that humans may be at risk. Animal studies provide one of two complementary approaches to the assessment of carcinogens, the other being epidemiologic studies.

Use of Animal Toxicity Studies

Animal studies are used to predict human risk. However, these studies are complicated by the fact that traditional methods of toxicology were not designed to deal with the pathogenic mechanisms involved in the chemical induction of cancer, since the appearance of carcinogenic effects is delayed. We are dealing with a "new toxicology"; and, consequently, a number of new methods have been proposed to analyze dose-response relationships for cancer and to extrapolate findings from animals to humans. As knowledge of the biology of cancer has expanded, new mathematical models have been developed—no single model being able to satisfy all of the scientific community.

The fact that outcomes observed in laboratory animals are applicable to human populations (if the findings are appropriately qualified) [2] is an important operating assumption. Nearly every form of human cancer has an experimental counterpart. There is a considerable body of data to convince us that those substances that are carcinogenic in animals are also carcinogenic in humans, and vice versa. With the possible exceptions of arsenic and benzene, all human carcinogens can be modeled in at least one animal species.

The use of long-term animal tests rests on the principle that the administration of high doses represents a valid method of discovering possible human hazards. The results of high-dose tests in animals are often criticized as being not relevant to human risk; regardless, this practice is justified because of the small number of animals that can feasibly be studied and the low probability of a tumorigenic outcome. The problem of validity has been addressed by several models for estimating risk at doses that are typically lower than the experimental dose. We recognize that even the most conservative approaches may underestimate risks, since experimental assays are conducted with genetically homogeneous test species under well-controlled environmental conditions. In contrast, human populations are genetically heterogeneous and live under a variety of dietary and environmental conditions.

The problem of extrapolation from laboratory animals to the human environment is also a concern. The evaluation of species-to-species differences in response to chemical carcinogens is a problem of comparative pharmacology [3]. Aspects of absorption, distribution, metabolism and/or excretion processes must be evaluated on a species-to-species basis, before an overall judgment of predictability can be made.

One of the several specific considerations relates to excretory rates. Small animals excrete compounds faster than large ones in a systematic, and therefore, statistically controllable way. Temporal differences are a second consideration. Cell division rates are more rapid in smaller animals, while the latent period may be longer in large animals. Life spans differ; for example, the human life span is about 35 times longer than that of a mouse or rat. A third consideration is body size, which affects the rate of distribution of a foreign compound throughout the body. In a mouse, cardiac output per minute is equal to its blood volume, about 1 ml/min. In humans, cardiac output per minute is only 5% of the blood volume. Therefore, the mouse distributes whatever is in the blood about 20 times faster. Another aspect of body size is that large animals have far greater numbers of susceptible cells that may interact with potential carcinogenic agents; this is important since there may be a relationship between the initiation of a carcinogenic event and the number of eligible cells on which a carcinogen could act.

The kinetics of a chemical may actually be dose-dependent. If metabolic and excretory processes are saturable, large doses may overwhelm these processes, and there will be a disproportionate increase in tissue concentration and consequent toxicity [4]. The dose-dependent fate of many chemicals, including aniline, benzopyrene, salicyclamide and styrene, has been demonstrated empirically, and this is important in interpreting extrapolation from high to low dose.

Despite the many qualifications that must be placed on animal assays as predictors of human carcinogenicity, these tests are an integral part of risk assessment. Results of the assays often lead to testable hypotheses in human populations or assist in confirming observations first made on humans. The value of animal studies will increase substantially as more sensitive techniques are developed for the evaluation of responses to low doses.

Use of Epidemiologic Studies

The complement to animal tests in carcinogen assessment is the epidemiologic study. Analyses of large human populations, often involving international comparisons, have provided presumptive evidence that differences in cancer occurrence are attributable to environmental factors. Within the United States, for example, substantial regional variation exists in cancer incidence and mortality patterns [5]. Studies of geographic differences are a powerful tool in generating etiologic hypotheses and identifying high-risk groups. While aggregate population studies cannot in themselves confirm causal or dose-response relationships, they offer the advantage of a relatively easy and inexpensive way to test preliminary general hypotheses, refine them for studies of individual cases and controls, and provide the first step in a link between epidemiologic and laboratory testing results.

An even more powerful method in analytic epidemiology is the case-control study. Basically, this design compares the rate of disease in an exposed group with the rate in an unexposed group. From such data, the risk of exposure can be evaluated quantitatively.

There are limitations to the epidemiologic approach. Modest environmental effects may not be detected, often because of inadequate measures of exposure. It is also difficult to measure the effect of a single exposure; the environment is complex, and factors confounding the exposure can make it difficult to control extraneous effects either by research design or statistical analysis. Despite these difficulties, the epidemiologic approach is well-warranted. In many cases, such studies provide sufficient information for development of control strategies even before the underlying biologic mechanisms have been determined. Epidemiologic studies are also invaluable in assessing risk from agents that are not harmful to animals or for which appropriate animal models have not yet been derived. Another advantage is that epidemiologic studies are not limited by the uncertainties of interspecies variation.

Other Problems

Two perplexing problems cannot be dealt with adequately by either laboratory or epidemiologic models. On theoretical and empirical grounds, the interactive, synergistic or antagonistic effects of carcinogenic substances bear on their health implications and subsequent control. There are many examples of synergistic effects in the pathogenesis of cancer. Some agents may potentiate the effect of others, even though they are not themselves carcinogenic. For example, sulfur dioxide and benzo[a]-pyrene (B[a]P) together yield respiratory tract tumors in hamsters and rats; yet neither agent alone produces tumors [6]. It is also well established in human populations that cigarette smoking disproportionately increases the risk of lung cancer in persons occupationally exposed to asbestos [7].

Neither animal nor human models can adequately represent the true complexities of the environment. Tests of single compounds in controlled laboratory settings may not identify crucial interactions that occur in the human environment. On the other hand, the natural setting of epidemiologic investigations does not generally allow for control, or even identification, of all relevant confounding factors. These and similar findings suggest that research models are needed that better reflect the environmental conditions under which heterogeneous human populations live. More powerful statistical models are also necessary to help explicate nonlinear (e.g., synergistic) relationships between dose and response [8].

Second is the problem of latency. Clinical disease may not be apparent for many years after the onset of the relevant exposures. Thus, the link between exposure and outcome is often difficult to make because the exact length of the latent period is unknown. This is further complicated because of the lack of historical exposure data. Animal studies generally do not help because of the inadequate experimental techniques that attempt to compensate for life-span differences. Also, only a handful of human epidemiologic data adequately describe the temporal relationship between exposure and outcome.

The importance of understanding latency in the examination of cancer rates is most dramatically demonstrated in the case of smoking and female lung cancer. Over the last 20 years, lung cancer rates have increased more rapidly for females than for males. There is little doubt that this increase is a consequence of the change in female smoking habits. The prevalence of smoking among females began to increase in the 1930s and

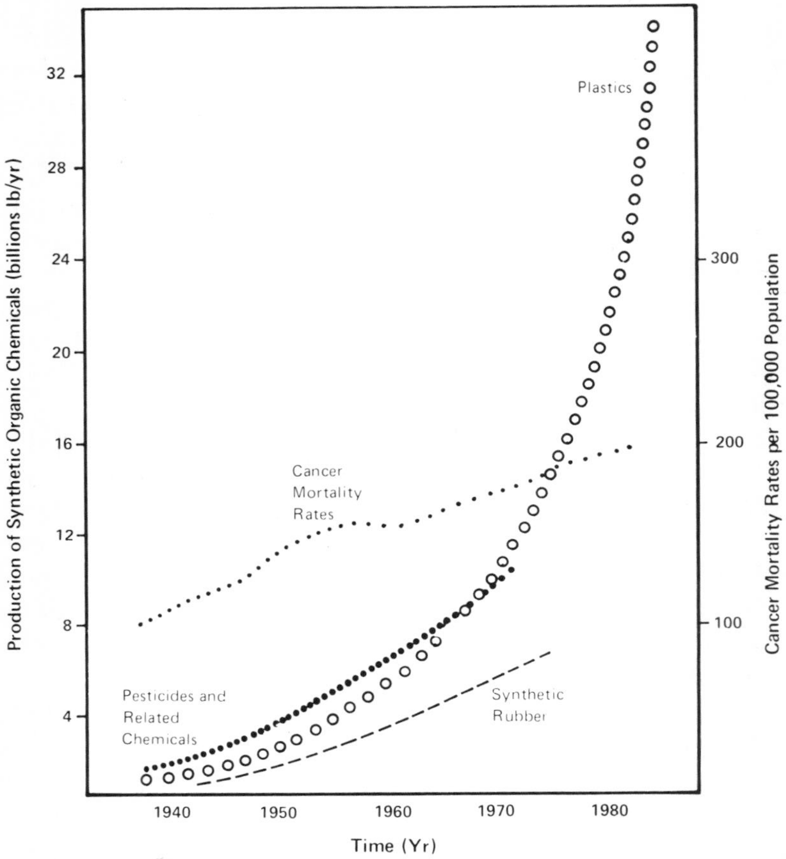

Figure 1. Cancer mortality rates and chemical production as a function of time.

peaked as women entered the work force during World War II. An increase in female lung cancer mortality rates began about 1965, reflecting the end of the latency period for the tumors induced by the earliest exposure [9]. The identification of the latent period for these women can be defined even better by examining age-specific mortality rates. Between the periods of 1956–1958 and 1974–1976, the greatest increases occurred for women of ages 50–65 years, representing cohorts born between 1910 and 1925. These women began smoking as they reached adulthood, thereby being at greater risk by age 50 than earlier birth cohorts, who did not adopt the habit until their mid-30s.

The temporal relationship between exposure and outcome is usually more subtle. The current increase in cancer rates probably reflects environmental exposures that began many years ago. This presumed relationship between new chemicals in the environment and cancer rates is expressed in Figure 1. Note that cancer mortality rates have climbed rapidly, correlating with the growth in post–World War II synthetic chemical production. Alarmingly, the modest exposures prevalent today might be manifested as larger effects in the future as the latent period ends for these exposures.

It is important to understand temporal variables, so that changes in disease incidence can be evaluated in an appropriate exposure context. The implication is that the success of various interventions depends on their timing in addition to their validity. Thus, it is increasingly important to construct exposure models for epidemiologic studies that reflect an appropriate interval between onset of relevant exposure and outcome. This requires an understanding not only of the latent period, but the intensity of exposure over time and the translation of exposure into an effective dose rate.

EVALUATION OF CARCINOGENIC SUBSTANCES IN DRINKING WATER

This section presents, in some detail, the epidemiologic and related issues of current scientific and regulatory concern over carcinogens in drinking water. The rise of organic contaminants in drinking water provides a good example of the problem of assessing risks to humans from low-level, long-term exposures. The Safe Drinking Water Act of 1964 mandates a new set of policy determinations to achieve the health objectives set forth in the act. However, the scientific basis for limiting exposure of large populations to organic chemical compounds in potable water supplies is complex and not adequately resolved. What, then, is the historical perspective for the argument that certain constituents of drinking water may be hazardous? What recent studies provide necessary information? What are the limitations and qualifications that must be placed on evidence offered in support of that argument?

Historical Perspective

The drinking water controversy predates the first public report of possible human health effects. In the 1950s polycyclic hydrocarbons, such as 3,4-benzpyrene, were observed in water and recognized as a possible public health hazard. In the 1960s odor problems related to organic industrial pollution of the Kanawha River in West Virginia led to studies that identified additional organic contaminants [10]. However, 3,4-benzpyrene remained the most widely studied carcinogen, since it was detectable not only in drinking water, but in a variety of combustion products, including smoke and automobile exhaust. From the mid-1960s on, attention was focused on chemical constituents of finished, or treated, water, and thus the first class of evidence in the drinking water question—confirmation that potentially dangerous exposures do indeed exist. This, however, was only the first part of determining whether there is measurable risk to human health. The accumulated evidence from toxicologic studies and epidemiologic investigations was most relevant.

Subsequently, in the late 1960s through the 1970s, many studies confirmed the presence of other organic compounds in treated drinking water supplies in the United States and Europe. For example, in 1969 Schafer et al. [11] studied 10 municipal supplies drawing raw water from the Missouri or Mississippi Rivers. Although their main interest was in organochlorine pesticides, their metabolites and related compounds, they found dieldrin in more than 40% of the finished water samples, endrin, DDE and DDT in more than 30%, chlordane in 20%, and aldrin, heptachlor and hexachloroethane only occasionally.

An analysis in Evansville, IN, revealed more than 40 compounds, the major one being bis(2-chloroisopropyl)ether which was traced to an industrial outfall about 150 miles upstream [12]. A total of 32 contaminants were detected in Washington, DC, supplies, including a number of aliphatic and aromatic hydrocarbons, alcohols, brominated compounds and ketones [13]. By 1970 the total list of organic compounds found in finished water had risen to more than 300, and this number accounts for only about 10% of the total organic material.

The problem became more complicated when the source of many of these compounds was revealed. A large number of the compounds with suspected health effects were actually by-products of the very process used to disinfect water, namely, chlorination. Typical water treatment procedures result in the formation of trihalomethanes (THM), or haloforms, and related chlorinated compounds. Chlorination of naturally occurring humic substances in raw water accounts for the organohalogen by-products. Brominated products appear because the chlorine used for disinfection can oxidize natural bromides to bromine, a more reactive form [14].

The chemical mechanism that introduces THM into our water supplies is based on a classic haloform reaction, in which halogen combines with a simple methyl ketone or with a compound convertible to form a ketone. In water the most likely precusor material is natural organic matter, namely, aquatic humic or fulvic acids. The reaction itself is a function of the nature and concentration of the organic precursor, the nature of the aqueous chlorine species, pH, temperature and chlorine/carbon ratio.

Current Approaches

This expanding evidence of the organic contamination of drinking water was of concern. The U.S. Environmental Protection Agency (EPA) was mandated by the Safe Drinking Water Act (SDWA) to "conduct a comprehensive study of public water supplies and drinking water sources to determine the nature, extent, sources of and means of control of contamination by chemicals or other substances suspected of being carcinogenic" [15]. The National Organics Reconnaissance Survey was initiated to measure the concentrations of six volatile and halogenated compounds in the raw and finished water from 80 cities in the United States. Data for 83 additional cities were obtained by a similar survey in EPA Region V, which includes six Midwestern states.

Typical data are provided in Table I. Note that chloroform occurs at concentrations far exceeding levels of the other compounds, particularly 1,2-dichloroethane and carbon tetrachloride, which are of industrial origin, not chlorination by-products. Also, the levels of these contaminants in surface water generally exceed those in ground, or deep well, water—again a chlorination related finding.

Table I. Results of National Organics Reconnaissance Survey for Finished Water (Median Values Are Expressed in μg/1)

Compound		Water Sources	
Common Name	Chemical Notation	Surface	Ground
Chloroform	$CHCl_3$	29.019	6.000
Bromodichloromethane	$CHBrCl_2$	8.029	3.400
Dibromochloromethane	$CHBr_2Cl$	1.250	2.020
Bromoform	$CHBr_3$	0.207	0.988
1,2-Dichloroethane	$C_2H_4Cl_2$	0.470	0.280
Carbon Tetrachloride	CCl_4	0.502	0.867

Regardless of the accumulated evidence of the widespread contamination of drinking water, the actual risk to human health has not been determined. Recently, the Safe Drinking Water Committee of the National Academy of Sciences reviewed the toxicity of 309 volatile organic contaminants found in water. Of these, only 22 are known or suspected carcinogens [16]. For most of the 309, adequate chronic toxicity data are altogether lacking. Attention keeps returning, however, to chloroform, the toxic potential of which has been known for decades. As early as 1945, Eschenbrenner and Miller [17] gave oral doses of chloroform to mice over a four-month period. At doses high enough to produce liver necrosis (about 0.4 mg/kg body weight), hepatomas were observed only in the male mice. None were observed at the lowest dose levels.

Confirmation of the carcinogenicity of chloroform was provided by the 1976 bioassay conducted by the National Cancer Institute [18]. Chloroform was orally administered to mice in varying doses for 78 weeks. The animals were sacrificed at 92 or 93 weeks, and hepatocellular carcinomas were observed in all groups. In rats, the chloroform produced renal epithelial tumors. These and other studies leave little doubt that chloroform is a potent animal carcinogen. However, its risk to humans remains unclear; consequently, a number of epidemiologic investigations have been undertaken to examine this question.

A study released late in 1974 [19] by the Environmental Defense Fund sparked initial interest in further investigation of carcinogenic responses to drinking water. This investigation suggested that trace amounts of suspect organic chemicals in Mississippi River drinking water contributed to the high cancer mortality rates in Louisiana, particularly in New Orleans. The study used unweighted multiple regression analyses. Independent variables, on a county level, included degree of urbanization, median family income, occupation, population density and drinking water source. The water variable was defined as the percentage of persons receiving drinking water from the Mississippi River. Age-adjusted mortality rates for cancers of the gastrointestinal and urinary tract organs were treated as dependent variables, and several statistically significant regression coefficients were reported for the water variable.

Subsequently, other studies have been conducted using similar qualitative exposure variables, including 2 for the 88 counties in Ohio. One investigation compared cancer mortality rates between counties classified as either surface- or groundwater counties, depending on the water source for most of the population. The data were adjusted for urbanization, median income and percentage of the male population engaged in manufacturing. The mortality rates in surface-water counties were significantly higher than those in groundwater counties for cancers of the stomach and bladder in white males and for cancer of the stomach in white females [20].

In the second study, in Ohio, Harris et al. [21] computed unweighted multiple regressions, using the percentage of the county population receiving surface water as a continuous exposure variable. In males, regression coefficients were significant for cancers of the pancreas, bladder and esophagus.

Several studies attempted to refine the somewhat crude qualitative exposure variable by using quantitative chemical data of the National Organics Reconnaissance Survey. Three different statistical approaches were employed, although the basic data for the studies varied little. In my own analysis, unweighted hierarchical stepwise regressions were calculated for nine site-specific neoplasms of the gastrointestinal and urinary tracts [22]. Demographic variables were entered on the first step, and the water parameters were entered subsequently. In general, the findings failed to demonstrate consistently significant effects. The regression coefficient for chloroform was significant only for carcinoma of the pancreas among white males, but the importance of this finding is substantially reduced by the large number of statistical tests conducted. A second approach was used by Cantor et al. [23]. He weighted the mortality rates by the square root of the county population size, thereby allowing large counties to contribute more to the analysis. The results showed a consistent positive association between total THM content and male bladder cancer, although the finding is not highly significant. In a third approach, Hogan et al. [24] used a somewhat different weighting scheme, based on a maximum likelihood estimate. The most consistent associations were found for cancers of the colon and bladder.

Interpretive Problems

These and other studies suggest that the validity of the epidemiologic models used must be carefully evaluated [25].

While there are a number of complexities in the epidemiologic modeling of an ambient exposure I shall focus exclusively on issues related to the drinking water exposure variable itself.

Both qualitative and quantitative variables have been examined. Qualitative measures are indirect, and attempt to reflect known or hypothesized differences between various types of water sources or treatment practices. For example, the surface/groundwater dichotomy presumes that surface supplies contain more organic precursors and are more likely to be chlorinated. Another dichotomous variable, chlorinated vs unchlorinated water, is based on the empirical finding that chlorinated supplies are likely to contain higher concentrations of chloroform and other suspect THM. The predictive power of these variables is compromised, however, when surface or chlorinated supplies do not exhibit THM concentrations significantly higher than those of some ground or unchlorinated supplies. The converse may also be true. Relatively high levels of chloroform have been measured in supplies of large cities such as Dayton and Tucson, which rely on groundwater. The most outstanding example is Miami, FL, where groundwater had a chloroform concentration of 311 $\mu g/l$, the highest recorded value in the National Organics Reconnaissance Survey.

Quantitative measurements of drinking water quality also have some serious limitations, at least in the forms in which they have been used. First, only a small number of the compounds that have been identified have been analyzed systematically. Our working assumption that chloroform is an appropriate index compound may be incorrect. Second, the reported studies relied on samples taken at a single point in the distribution system and at a single point in time. However, the concentration of these compounds is known to vary seasonally and is affected by changes in pH and chlorine dose. Also, exposure may not be constant across the distribution system, since the haloform reaction continues as travel time from the treatment plant to the tap increases.

A satisfactory method for dealing with the problem of latency between initiation of exposure and evidence of disease has not been developed. However, our research group is currently investigating methods of using routine water treatment plant records as a way to retrospectively construct historical exposure models, and, therefore, provide an appropriate temporal sequence in the epidemiologic model.

The high mobility of the American population is another problem that can lead to misclassification errors. Only by interviewing individuals can detailed residential histories be obtained. This, however, is a complex and expensive undertaking. Individual variation in water consumption is another problem; this too would be difficult to determine, even by interview.

Even if the exposure variables were better understood, several technical problems regarding study design and statistical analysis would remain. The majority of the studies rely totally or, in part, on aggregate, or summary data rather than on individual observations. Such county-level data may obscure the heterogeneity of any characteristics measured. For example, Franklin County, OH, is not uniformly urban, although it appears to be so in a multiple-regression equation. Because these variables cannot be controlled as carefully as in individual studies, specific risk estimates cannot be derived from them. However, aggregate studies are important tools in epidemiologic reasoning; they are fruitful approaches for generating hypotheses, since they utilize readily available existing data and require limited fiscal resources.

If drinking water poses a threat to human health, the threat is a modest one. So, the question, "how powerful are our statistical techniques for detecting a modest effect on an already rare outcome?" becomes important. A crude, but reasonable, analysis suggests that the statistical power of a regression (that is, the probability of detecting an effect if it indeed exists) is very low [21]. By constructing a simple binomial model, and then varying the rate of cancer incidence due to effects other than water and/or due to water, statistical power can be estimated. Curve a in Figure 2 is a power curve with the background rate held constant at 30/100,000 and the water effect varying from 1×10^{-5} to 20×10^{-5}. Curve b assumes a 20% water effect, varying proportionately with the background ef-

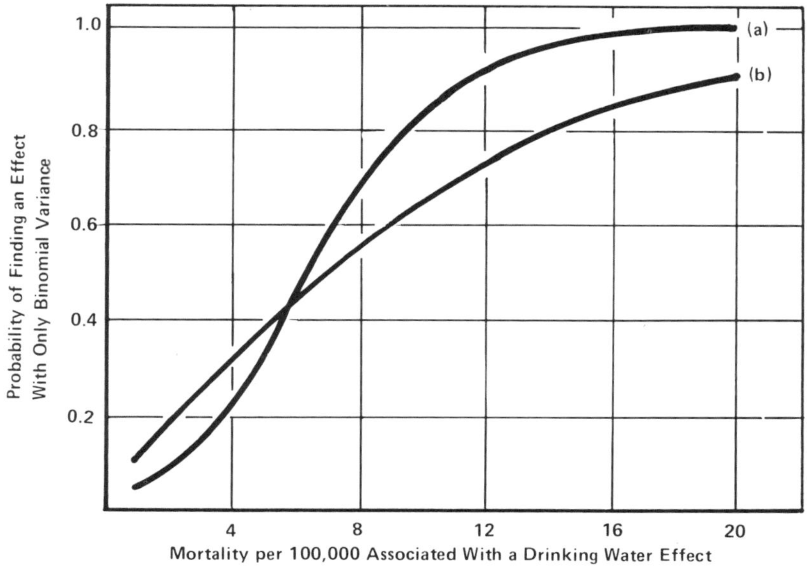

Figure 2. Statistical power as a function of the background cancer rate and the drinking water effect.

fect. Clearly, the water effect is more difficult to detect when it is small and/or if the cancer is rarer. I am concerned that our statistical methodology may be close to the limit of its capability to find such modest effects, particularly in light of all the noise in the system.

Even though these studies have failed to confirm unequivocally a drinking water effect, there still is justification for pursuing the drinking water question. The emerging consistencies, particularly with respect to cancers of the large intestine and bladder, may provide direction for future investigation.

ISSUES IN THE CONTROL OF TOXIC SUBSTANCES

Whether the issue is drinking water or another part of the environment, we ultimately confront questions about appropriate and prudent control measures. The identification of a negative health effect by epidemiologic observation often provides the necessary basis for intervention in the disease process even before sufficient knowledge about the exact biological mechanisms of action is available. Perhaps the most elegant example of this type of approach is the action taken by Snow in London in the midst of the 1854 cholera epidemic. After observing the pattern of the epidemic, Snow was convinced that cholera was somehow transmitted through water. Therefore, he removed the handle of the Broad Street pump, and the epidemic soon ended. Although we now know that the epidemic was already subsiding, this does not diminish the beauty of Snow's insight. He contributed greatly both by his action and by the suggestions he offered that subsequently led to the positive identification of the cholera vibrio. We are still somewhat in the same position as Snow was. Only today, the situation is far more complicated, and we do not always know which "pump handle" to remove.

It is when we develop intervention strategies that we need to reconcile the relative advantages and disadvantages of laboratory-experimental vs epidemiologic-observation techniques. That is, strategies for intervention must be designed and implemented in the face of often inconclusive, and sometimes contradictory, evidence. There are numerous historical instances of conflicting experimental and human evidence that resulted in delayed intervention. For example, workers who manufactured bis(chloromethyl)ether exhibited a cluster of lung cancer in 1962 [26]. It was not until 1971, however, that laboratory studies confirmed that neoplasms were induced in rodents by inhalation. However, serious attention had not been given to studies with other routes of administration.

The Regulatory Environment

In response to the growing concern over the toxic substances and carcinogenesis problem, more than 20 regulatory statutes have been enacted by Congress [27]. The enactment of this body of legislation reflects the assumption that the unregulated marketplace cannot solely be relied on to safeguard human health. These laws, while emphasizing the protection of human health, also recognize that we must strike a balance between health, economic and social goals. Consequently, many of the environmental laws entail some balancing of costs and benefits. However, this is not a consistent element, even among laws dealing with the same environmental factors. For example, the Federal Water Pollution Control Act (FWPCA) does not call for a balancing of costs; the SDWA, however, does.

Additionally, there are legislative inconsistencies in defining toxic substance. The FWPCA addresses contaminants that "will" cause disease or death, while the SDWA applies to substances that "may" have an adverse effect. This latter wording is clearly more precautionary and would result in limiting exposure to a "potentially" hazardous substance. Fortunately, lawmakers have begun to understand the inconclusive nature of many scientific findings and correspondingly are giving stronger expressions of the need for health protection.

The goal of primary prevention is most explicit in the Toxic Substances Control Act (TSCA), which requires the assessment of potential hazards before a chemical is manufactured for commercial use. This represents a desirable change from ad hoc, after-the-fact attention to toxic substances, which has generally led to crisis-oriented rather than planned strategies of intervention. Many previous laws, including the Clean Air Act, the Consumer Product Safety Act, and the Occupational Safety and Health Act, do not call for initiation of the regulatory process until a substance has been manufactured and has entered the environment.

Another advance in prevention may emerge from the recent emphasis by EPA on special toxicology testing for new chemicals that might be particularly toxic [28]. Each year, a priority list of 50 of the potentially most toxic new chemicals is to be established and subjected to extensive testing. The success of this approach depends on the criteria used for selecting chemicals for the priority list and the speed with which adequate tests can be conducted.

In addition to encouraging more preventive approaches, we must continue to confront the reality of environmental disease already induced. Epidemiologic methods can also play a key role in this regard. Increased emphasis should be placed on population surveillance to detect evidence of environmentally induced disease and related factors at the earliest opportunity.

The Need for Balance

Efforts must be intensified to define high-risk populations more thoroughly. Persons at risk should be characterized not only with respect to exposure status, but also with regard to age, race, sex, general health status or other host characteristics relevant to the agent in question. This will involve increased emphasis on the biological bases of human susceptibility, including genetic predispositions, immunologic sensitivities and nutritional deficiencies. An example of how such findings may help define high-risk individuals is that of aryl hydrocarbon hydroxylase (AHH) and lung cancer. Inducibility by AHH appears to be genetically controlled. In one investigation, smokers with intermediate and high AHH inducibility had 16 and 36 times more risk to lung cancer than those in the low group [29]. When properly confirmed, such findings could be translated into programs to identify high-risk individuals and screen them for early evidence of disease.

In addition to monitoring health outcomes of the host, proper exposure data—both occupational and ambient—must be collected. For example, industrial records should include information on specific exposures as well as follow-up data on the health of employees. If a hazard is detected, efforts to monitor the health of those exposed in the past can be intensified.

The evaluation and control of carcinogenic substances in the environment provides a significant challenge to epidemiologists, toxicologists, policymakers, and others. The continuing increase in production and use of chemicals adds to the difficulties of assessing the risk to human health. Although this discussion has focused on cancer as an endpoint, there is a broad range of other human health effects to be considered as well, including birth defects, heart and lung disease, and neurological disorders. It should be recognized that some substances, such as lead, have been associated with several effects, an observation that further complicates the issue.

At the heart of the issue is the balance between scientific testing and judicious regulation. Rarely is there consensus on the extent of risk or the approach to control. Scientific evaluations of risk are probabilistic in nature, and, therefore, intrinsically uncertain. Regulation, on the other hand, is deterministic in the sense that a regulatory decision is specific, e.g., ban or fail to ban a substance, or set a maximum exposure level at an exact concentration.

These inherent differences between the scientific and regulatory decision-making modes further complicate the assessment and control of environmental hazards. While the essential conflict may never be resolved, progress can be made by the careful execution and reporting of scientific investigations. Within the limits of probabilistic decision-making, scientific input can be strengthened by developing experimental and observational models that adequately reflect the underlying biologic assumptions and at the same time limit the number of confounding influences. Similarly, regulatory agencies must be increasingly cognizant of the strengths, as well as the limitations, of scientific procedures and inference. The scientific and regulatory communities must attempt to strike a mutually satisfactory balance between their respective decision-making strategies to ensure maximum protection of human health.

REFERENCES

1. Pollack, E. S., and J. M. Horm. "Trends in Cancer Incidence and Mortality, 1969–1976," *J. Nat. Cancer Inst.* 64:1091–1099 (1980).

2. Schneiderman, M. A., and C. C. Brown. "Estimating Cancer Risks to a Population," *Environ. Health Pers.* 22:115–124 (1978).
3. Rall, D. P. "Species Differences in Carcinogenesis Testing," in *Origins of Human Cancer*, H. H. Hiatt, J. D. Watson and J. A. Winsten, Eds. (Cold Spring Harbor, NY: Cold Spring Harbor Laboratory, 1977), pp. 1383–1890.
4. Levy, A. "Dose-Dependent Effects in Pharmacokinetics," in *Importance of Fundamental Principles in Drug Evaluation*, D. H. Tedeschi and R. E. Tedeschi, Eds. (New York: Raven Press, 1968).
5. Mason, T. J., F. W. McKay, R. Hoover, W. J. Blot and J. P. Fraumeni. "Atlas of Cancer Mortality for U. S. Counties: 1950–1969," DHEW Pub. No. (NIH) 75–780 (1975).
6. Kuschner, M. "The Causes of Lung Cancer," *Am. Rev. Resp. Dis.* 98:573–590 (1968).
7. Selikoff, I. J., H. C. Hammond and J. Chung. "Asbestos Exposure, Smoking, and Neoplasia," *J. Am. Med. Assoc.* 204:106–112 (1968).
8. Rothman, K. J. "Estimation of Synergy or Antagonism," *Am. J. Epidemiol.* 103:506–511 (1976).
9. Burbank, F. "U. S. Lung Cancer Death Rates Begin to Rise Proportionately More Rapidly for Females than for Males: A Dose-Response Effect?" *J. Chron. Dis.* 24:475–479 (1972).
10. Rosen, A. A., R. T. Skeel and M. B. Ettinger. "Relationship of River Water Odor to Specific Organic Contaminants," *J. Water Poll. Control Fed.* 35:777–782 (1963).
11. Schafer, M. L., J. T. Peeler and W. S. Gardner. "Pesticides in Drinking Water: Waters from the Mississippi and Missouri Rivers," *Environ. Sci. Technol.* 3:1261–1269 (1969).
12. Kloepfer, R. D. "Characterization of Organic Components in a Municipal Water Supply," *Environ. Sci. Technol.* 6:1036–1037 (1972).
13. Scheiman, M. A., R. A. Saunders and E. E. Saalfeld. "Organic Contaminants in the District of Columbia Water Supply," *Biomed. Mass Spectrom.* 1:209–211 (1974).
14. Rook, J. J. "Formation of Haloforms During Chlorination of Natural Waters," *Water Treat. Exam.* 23:234–243 (1974).
15. Public Law 93-523, "Safe Drinking Water Act," *Federal Register* 40:11990–11998 (1975).
16. Safe Drinking Water Committee, National Research Council.

Drinking Water and Health (Washington, DC: National Academy of Sciences Printing and Publishing Office, 1977).
17. Eschenbrenner, A. B., and E. Miller. "Induction of Hepatomas in Mice by Repeated Oral Administration of Chloroform, with Observations on Sex Differences," *J. Nat. Cancer Inst.* 5:251–255 (1945).
18. Page, N. P., and U. Saffiotti. "Report on Carcinogenesis Bioassay of Chloroform," National Cancer Institute, Division of Cancer Cause and Prevention. Bethesda, MD (1976).
19. Harris, R. H. "Implications of Cancer Causing Substances in Mississippi River Water," Environmental Defense Fund, Washington, DC (1974).
20. Kuzma, R. J., C. M. Kuzma and C. R. Buncher. "Ohio Drinking Water Source and Cancer Rates," *Am. J. Public Health* 67:725–729 (1977).
21. Harris, R. H., T. Page and N. A. Reiches. "Carcinogenic Hazards of Organic Chemicals in Drinking Water," in *Origins of Human Cancer*, H. H. Hiatt, J. D. Watson and J. A. Winsten, Eds. (Cold Spring Harbor, NY: Cold Spring Harbor Laboratory, 1977).
22. Reiches, N. A. "An Epidemiologic Investigation of the Relationship Between Chemical Contaminants in Drinking Water and Cancer Mortality," PhD Thesis, The Ohio State University, Columbus, OH (1977).
23. Cantor, K. P., R. Hoover and T. J. Mason. "Association of Cancer Mortality with Halomethanes in Drinking Water," *J. Nat. Cancer Inst.* 61:979–985 (1978).
24. Hogan, M. D., P. Chi and D. G. Hoel. "Association Between Chloroform Levels in Finished Drinking Water Supplies and Various Site-Specific Cancer Mortality Rates," *J. Environ. Pathol. Toxicol.* 2:873–887 (1979).
25. Wilkins, J. R., N. A. Reiches and C. W. Kruse "Organic Chemiical Contaminants in Drinking Water and Cancer," *Am. J. Epidemiol.* 110:420–448 (1979).
26. Weiss, W., and K. R. Boucot. "The Respiratory Effects of Chloromethyl Methyl Ether," *J. Am. Med. Assoc.* 234:1139–1144 (1975).
27. Toxic Substances Strategy Committee. "Toxic Chemicals and Public Protection, A Report to the President," U.S. Government Printing Office (1980).

28. Olsen, K. "Review of Federal Toxic Substances Legislation," paper presented at the New England Air Pollution Control Association Meeting, Cambridge, MA (1976).
29. Kellerman, G., C. R. Shaw and M. Luyten-Kellerman. "Aryl Hydrocarbon Hydroxylase Inducibility and Bronchogenic Carcinoma," *New England J. Med.* 289:934–937 (1973).

4

Ecosystem Theory and the Unexpected: Implications for Environmental Toxicology

Robert V. O'Neill and Jack B. Waide
Environmental Sciences Division
Oak Ridge National Laboratory
Oak Ridge, Tennessee

In the twentieth century, Western society has had a deplorable record of misjudging responses of environmental systems to potentially toxic substances. Chlorinated hydrocarbons, DDT and mercury are well-known examples of substances that affected ecosystems in unanticipated ways, bearing out our experience that the responses of ecosystems to human intervention are not easily predicted. At times, these responses even surprise us.

Much of the toxic substance problem has resulted from gross negligence in application and disposal of toxicants. Part of the problem, however, is due to our failure to understand ecosystem dynamics. The negligence may be controllable by regulation, but the ignorance can only be corrected by significant changes in the way we view environmental systems and conduct toxic substances research.

Toxicological testing protocols usually involve single-species systems such as rodents or fish, and are designed to provide quick and unequivocal results. Basic research in toxicology typically focuses on detailed physiological mechanisms involved in the toxic or carcinogenic response. The resulting information is invaluable, and research in these areas should be continued and intensified. At the same time, this is inappropriate information for predicting effects on entire ecosystems.

In this chapter, we focus exclusively on ecosystem-level effects. True, in many cases, effects on components, e.g., humans or a sensitive species, may provide the most relevant information and represent the most practical focus for regulatory purposes, but there is another important aspect of the problem—response of the total system. Thus, we deal with only one part of the toxic substance question, i.e., what concepts and measurements would be useful for predicting ecosystem-level effects? We are neither maintaining that this is the only question that needs to be addressed, nor that for any specific chemical it is necessarily the most important. It is also clear that the current state-of-the-art of ecosystem analysis is not adequate for the task. At best we can offer suggestions and introduce concepts for discussion. Nevertheless, the question is an important one. Anthropogenic chemicals are continuously being released into intact ecosystems. Therefore, it is important to improve our understanding of integrated system responses to predict long-term and often subtle responses of the ecosystem, which are not predictable from current emphases on single-species responses and human dose rates from epidemiologic studies.

The ecosystem has long been recognized as a fundamental organizational unit in ecology [1–4]. Yet, ecologists have been slow to recognize or identify holistic behaviors at the ecosystem level of organization [5]. Most ecologists consider the ecosystem as the matrix or environmental context within which the population or community operates. The focus, however, is on the component, and not the system. Fewer ecologists approach the ecosystem as an integrated system with measurable properties that transcend any simple summation of component processes [6–11]. Yet, it is precisely this perspective that is required for understanding and predicting ecosystem responses to toxic substances.

A major stumbling block in the study of ecosystems is their bewildering complexity. Like other natural systems, ecosystems are complex systems "made up of a large number of parts that interact in a non-simple way" [12]. Even if it were possible to elucidate "the properties of the parts and the laws of their interaction, it is not a trivial matter to infer the properties of the whole" [12]. Yet, complexity of organization would seem to be a key element in the homeostatic response of ecosystems to external disturbances [13]. Thus, understanding ecosystem complexity may be essential to management of the consequences of human interventions.

Ecosystems are among the so-called "middle-number systems" [14,15]. In such systems, there are not enough components to average their behaviors reliably, yet there are too many components to write a single equation for each. Thus, many of the analytical approaches that have proven so useful in physical or engineering sciences are less useful for examining ecosystems. Indeed, Waide and Webster [16] suggested that, since the "design constraints" that have guided the evolution of these systems are not well understood, the large array of techniques developed for systems engineering may be of lesser use when applied to ecosystem investigations. Hill and Durham [17] similarly outlined a variety of assumptions implicit in control theory, but which may not apply in ecology. Rosen [18] detailed other objections to the use of classical systems approaches for studying biological systems.

Faced with this complexity, ecologists have historically taken a more component-oriented approach to ecosystems. However, any approach to the study of ecosystems that focuses on components is deficient because it ignores the overall functioning of the system. The ecosystem, like any biological unit, is a thermodynamically open, dissipative system [19] in which coupled anabolic and catabolic processes permit the formation of persistent organic structures out of a surrounding inert geochemical matrix. Failure to recognize ecosystems as integrated, biogeochemical systems represents a major conceptual barrier to the development of a functional theory of the ecosystem, as well as to the application of any such theory to practical problems, including toxic substances, of ecosystem management.

The essence of the problem inherent in the component-oriented approach may be stated quite simply: components behave very differently when isolated from the ecosystem of which they are a part. Measuring properties of isolated components provides little, if any, information about the behavior of these components as functioning parts of natural ecosystems. Nor can we infer behavior or properties of the whole ecosystem from measurements on components isolated from the ecosystem. Weiss [20] has perhaps stated the point most clearly: "The state of the whole must be known in order to understand the collective behavior of the parts." Schindler et al. [10] reiterated this point in the recommendation:

> The attempt to understand the "state of the whole" should thus be a central focus in all ecosystem studies. Only after this state has been assessed or measured should we attempt to fractionate the system into component subsystems. Indeed, it should be our knowledge of the "state of the whole" which guides our definition of subsystems, rather than a prior definition of subsystems delimiting our recognition of properties of the whole.

Certainly a major focus of toxic substance programs should be understanding effects on the "state of the whole" ecosystem in addition to understanding effects on human and other populations.

Our intent in writing this chapter is to defend, illustrate and amplify on the points stated above, as they relate to the management of toxic discharges. In the following section we develop the notion of system as it emerges from hierarchy theory. The next section emphasizes the ecosystem as a biogeochemical system. Next, we focus explicitly on the system-component problem and provide examples of ecological problems involved in extrapolating from component measurements to system behaviors. Finally, we make some recommendations for toxic substances research that emphasize an integrative approach to the measurement of system as opposed to component behavior.

ECOSYSTEMS AS HIERARCHICAL SYSTEMS

Any predictive theory of ecosystem dynamics must depend on a thorough appreciation of the ecosystem as a functionally integrated metabolic system and the proper relationship between system and component properties. Yet, for "middle-number systems" such as ecosystems [14] obtaining such understanding may be especially difficult. Is there a way? Allen and Starr [15] argued that general systems and hierarchy theories represent the only working models or metatheories [21] we have at our disposal for better understanding complex systems such as ecosystems.

Rate constants associated with various levels of system organization are central to the hierarchy concept. From a hierarchical perspective, the definition of system structure is more strongly related to the frequencies at which certain behaviors occur than to the exact nature of the behaviors themselves [12,22,23]. That is, the set of behaviors that may be defined for any given system may be arrayed according to the frequencies with which they occur. Behaviors corresponding to higher levels of system organization occur with lower frequencies (i.e., have longer time constants) and, conversely, lower levels of organization within the system exhibit higher-frequencies (i.e., have short time constants). For example, turnover of certain elements in the total system may involve decades or centuries while turnover of the same elements in a bacterial population can be measured in hours.

Typically, systems are conceptualized so that major attention is focused at a single organizational level. Behaviors attributable to higher levels occur very slowly and appear in the system description as constants. In contrast, lower-level behaviors occur so rapidly that only their average or steady-state properties are relevant. Simon [22] referred to this range of conditions as "loose vertical coupling" among hierarchical levels, and suggested that conceptualization of a given system is effective only if it permits a "sealing off" of higher and lower levels of behavior. The overall organization of the system cannot be expressed adequately by one frequency, it is "nearly decomposable" [22] because each level operates with distinctly different time constants (see also Patten et al. [24]).

Further structuring of hierarchical systems is based on interactions among components. Strongly interacting components form subsystems within any level of the hierarchy. Interactions among subsystems on the same level are comparatively weak. Simon [22] termed this within-level organization "loose horizontal coupling." For example, short-term dynamics of a subsystem are dominated by strong interactions among its components (e.g., individuals in a population). Interactions among subsystems (e.g., among populations in a community) involve weaker interactions, and occur at lower frequencies and with longer time constants, eventually to reach a system-level view.

In this way, weak interactions among subsystems at one level in the hierarchy define the dominant time constant or behavioral mode at the next higher level. Hence, differences in the strengths of interactions give rise to the structure of hierarchical systems and result in "a scale of decreasing behavioral frequencies corresponding to increasingly complex levels of system organization" [10]. As a consequence of this hierarchical organization, dynamics of a component are averaged, filtered or smoothed into the aggregate output for the subsystem of which that component is a part. It is the subsystem behavior which is seen at the next higher level in the hierarchical system. Further filtering occurs at each successive level. Thus, the observation that there is no clear or straightforward relationship between total system behavior and the properties of an isolated component is a direct consequence of the hierarchical view of natural systems.

The major challenge in studying the ecosystem is to make explicit the frequency-dependent organization inherent in system structure. Viewing the system at a finely resolved spatiotemporal scale may identify component responses that have little direct bearing on the long-time, large-space response of the intact ecosystem. Clearly, to understand system response, we must identify the proper spatiotemporal scale of observation. This scale of observation must correspond to the structural organization of the ecosystem, not of the observer, and thus may differ radically from the lifetime and perceptual abilities of humans [25]. In short, we must view ecosystems in the frequency domain in which the system operates, rather than the domain in which we ourselves, one of the components, or subsystems, operate.

THE ECOSYSTEM: A BIOGEOCHEMICAL SYSTEM

From a fundamental viewpoint, the ecosystem is a stable configuration of matter that persists, in a thermodynamic sense, by degrading energy more capable of doing chemical work to energy less capable of work. The ecosystem forms high-energy organic compounds out of low-energy inorganic compounds that are mobilized from the surrounding geochemical environment. Thus, the ecosystem is basically a biogeochemical system [10]. Such a viewpoint is consistent with the hierarchical approach outlined above in that it emphasizes ecosystems as integrated metabolic networks of interacting biological, physical and chemical processes existing within definable landscape units. Hutchinson [26] was one of the first to emphasize this biogeochemical nature of ecosystems in his discussion of "circular causal systems."

Energy in solar radiation is trapped by photosynthetic pigment assemblies in autotrophic components of ecosystems and is then converted into chemical bond energy. This chemical bond energy is degraded to build persistent, high-energy organic structures out of low-energy inorganic molecules taken up from the surrounding inorganic geochemical matrix. After senescence or death, organic structures are broken down, with the released energy being ultimately dissipated back to the environment as heat and the elements being returned to the geochemical matrix in a state of low chemical potential (although not necessarily all the way to purely elemental forms, such as C or N).

Element cycles are a necessary consequence of energy dissipation at the ecosystem level [27,28], and both element cycling and energy fixation-dissipation may be viewed as complementary and interrelated functional processes of ecosystems [29]. Thus, it is appropriate to view the ecosystem as a thermodynamically open, nonequilibrium, dissipative system that builds and maintains persistent organic structures out of an inorganic geochemical matrix. This overall process is achieved by coupling energy-absorbing (anabolic) and energy-releasing (catabolic) processes [10].

Such a biogeochemical perspective of ecosystems should prove to be an especially useful one for examining the behavior of toxic substances in ecosystems. It compels us to view the ecosystem not as a set of interacting species populations but as an intact metabolic network [30]. The perspective is especially attractive when it is realized that many toxic substances introduced into the natural environment mimic the behavior of elements naturally flowing along biogeochemical pathways [31–33]. Additionally, many toxic substances alter rate processes in ecosystems, thereby disrupting biogeochemical cycles in subtle, albeit detectable, fashions [34].

This biogeochemical perspective has been applied to both natural and stressed ecosystems. One of the earliest and most comprehensive uses of the approach was that of Baas-Becking et al. [35]. These authors borrowed the two "master variables" [36] of chemical equilibrium analyses, pH (–log proton activity) and pE (–log electron activity) to characterize the biogeochemical state of natural environments. (They actually measured pH and Eh, the latter variable being a more easily measured approximation of pE.) Measurement of these two variables allowed the construction of "boundary conditions" for specific types of autotrophic organisms (e.g., blue-green algae vs mobile green algae) within a two-dimensional pH-Eh phase space. Because pH and pE (or Eh) provide an integrated measure of coupled biochemical and geochemical processes, Baas-Becking et al. [35] provided a rationale for viewing the ecosystem as a complex but integrated system of redox reactions occurring within a specific geochemical matrix [10].

Following this same approach, Schindler et al. [10] and Waide et al. [37] characterized the dynamics of aquatic laboratory microcosms in a two-dimensional, pH–dissolved oxygen (DO) phase space. Again, DO was used as a reasonable surrogate of pE for the aerobic microcosms investigated. Distinct organizational states of these microcosms, independent of taxonomic composition, were identified as definable regions within the pH-DO phase space. The overall level of metabolism within the microcosms was related to the shifts in the microcosms between different organizational states as well as to the uptake and remineralization of critical elements such as phosphorus. Moreover, using this biogeochemical approach, Waide et al. [37] were able to measure whole-system responses to physical, chemical and biological perturbations. The respective perturbations were achieved by alteration of photoperiod and stirring regime, titration to acidic or basic conditions, and addition of longer-lived herbivores.

In terrestrial environments, one of the most prevalent biogeochemical approaches to the study of ecosystem dynamics has involved small watersheds as experimental study systems [38]. In these studies, net budgets (i.e., inputs minus outputs) of select elements are used to assess the state of the ecosystem. Such watershed studies, especially in forested landscapes, have proved useful for characterizing the baseline biogeochemistry of undisturbed ecosystems [39–41], for comparing biogeochemical responses of ecosystems located in diverse regions of this country [42,43] and for measuring responses of ecosystems to specific anthropogenic alterations [44–46]. This approach provides an integrated measure of total ecosystem behavior which can be related to processes of element recycling and conservation [43,47,48].

Finally, studies of terrestrial microcosms have shown that a biogeochemical perspective provides a system-level measure of response to toxic substances. O'Neill et al. [34] demonstrated that element loss in leachate is a sensitive total-system response to heavy metals. Similarly, Van Voris et al. [49] showed that spectral analyses of CO_2 efflux from intact soil cores was related to microcosm responses to cadmium. Together, these studies demonstrate that a biogeochemical perspective of ecosystem dynamics compels us to recognize the ecosystem as a functional metabolic system, and provides us with experimental means of assessing holistic ecosystem responses to the introduction of toxic substances.

SYSTEM-COMPONENT INTERRELATIONS

Having presented the hierarchical approach and pointed out the advantages of defining the ecosystem as a biogeochemical system, it remains to identify and draw out the implications of this approach to ecosystem theory. In particular, we want to emphasize further the differences between a system- and a component-oriented view of the ecosystem, and to argue that the ecosystem "surprises" us because our viewpoint is too limited. There are three principal points.

Ecosystem Constraints on Component Behaviors

In a seminal study, Gardner and Ashby [50] demonstrated theoretically that large, highly connected systems are unlikely to persist. Specifically, they showed that when a large number of components are randomly connected, the probability that the system will be unstable increases rapidly as the number of interactions increases. A key factor in their analysis was that interactions are random. In any large, stable system, such as an ecosystem, interactions are not random but are constrained to follow specific "rules" [51]. In ecosystems, these rules may take the form of density-dependent regulation of higher trophic levels, food limitation and inefficient energy conversion. DeAngelis [52] has shown that in simulated ecosystems which follow such nonrandom rules, the results of Gardner and Ashby must be modified substantially since stability can increase with increased connectance. Waide and Webster [16] showed comparable results with different analytical procedures.

The resulting interaction structure, or organization of rules, ensures the stability of the system, but achieves this result by imposing constraints on the dynamic behavior of the components. These constraints may change. A system characterized by a high degree of component interaction may, under certain circumstances, "collapse" according to Allen and Iltis [23] into a stable, high-order system. The collapse may involve the elimination of some species; however, over evolutionary time, most populations will persist by coevolving a stable configuration of interactions. A similar result has been suggested by Tregonning and Roberts [53]. As collapse to ever higher levels of organization proceeds over evolutionary time, with the concomitant achievement of higher organizational states, the constraints placed by the system on components become greater. Thus, evolutionary collapse to higher levels leads to "more explicit and fine-grained control of low-level parts by large-scale wholes" [23].

These constraints are evident at all levels of the biological hierarchy. For example, nerve cells—a component of the whole organism—growing in the animal body are constrained during development into very specific morphologies. Removed from the constraints of the body and grown in tissue culture, these cells show a tremendous increase in plasticity. Likewise, an ecosystem component (i.e., population or community), isolated from its contextual system, can show a much greater range of dynamic behavior than the same component in its natural context. As a trivial example, many populations growing under optimal laboratory conditions will show growth rates that cannot even be approximated in the field. At the other extreme, plants without their natural mycorrhizal components grow more slowly in the laboratory.

Less trivial examples of varying levels of constraints are provided by species introduction [54] and gnotobiotic microcosms [55]. When a species is transposed to a new system, the constraints evolved in its original context are removed. The new ecosystem contains limited food supplies, predators and competitors, but these have not specifically coevolved with the species. Under the latter circumstances, the introduced species may destabilize the system, showing dynamic behavior that has never been observed in its natural environment. Introduction of rabbits into Australia and chestnut blight into the eastern United States serve as graphic examples of ecosystem disruption. A similar phenomenon occurs when a few species are assembled into a simple microcosm. Typically, some or all of the populations in the microcosm show unconstrained behavior, and the system collapses. The artificial assemblage does not contain the same complex of biotic interactions developed over evolutionary time to constrain component behavior and to stabilize the system. Patten [56] has even argued that the necessity for ecosystem components to interact within definable environments "provides constraints that almost certainly guarantee coevolution to be an ecosystem level phenomenon."

The fact that organisms have evolved as interacting, constrained components of a structured system poses a serious dilemma for researchers investigating toxic effects on system components. The isolated organism or population is as much an artifact as the nerve cell in tissue culture. Measured changes in the organism's dynamic behavior (e.g., mortality or fecundity), may have little or no bearing on its behavior within its natural ecosystem.

Notice that we are not saying that it is difficult to design a properly controlled laboratory study. We are saying that it is impossible to duplicate the natural constraints. It may be possible to control the physical environment that constrains the organism, but it is not possible to control the complex of biotic interactions. In addition to their large number, these interactions occur on spatial and temporal scales that cannot be reproduced in the laboratory. There is nothing that one can do in the laboratory to guarantee that a toxic effect (or lack thereof) measured on an isolated component will be observed in the ecosystem.

In fact, if one had isolated measurements on every component of the system, there is still no guarantee that the behavior of the intact ecosystem could be reproduced. Free body modeling (e.g., Caswell et al. [57]) may be a theoretical ideal, but its practical application, especially in ecology, is questionable. At least that has been a common experience of modelers who have assembled isolated physiological- and population-level data. The resulting ecosystem model is capable of a broad spectrum of behavior, much of which is unstable and could not persist in the real world. This unrealistic behavior occurs because critical and constraining interactions are not adequately represented in the collected data and consequently are not incorporated in the model.

This is not to say that laboratory studies produce no useful information. Certainly a measurable toxic effect can provide input to management for use in controlling releases. This is different, however, from predicting that a toxic effect will or will not be seen in the ecosystem. We should not stop doing laboratory studies, but we should stop being surprised when the ecosystem does the unexpected. The critical importance of ecosystem organizational rules, not measurable in experiments of isolated components, is predictable from hierarchy theory. The result is lamentable, but not unpredictable.

Organizational Structure and Component Interactions

It is the evolution of an ecosystem structure that transforms the chaos of hundreds or thousands of interacting components into a stable ecosystem. To understand toxic effects on the ecosystem, therefore, we must shift our focus from the components to the interactions and to the organizational "rules" that structure these interactions. This shift places a new emphasis on indirect or higher-order effects occurring in the system.

In any complex system, the activities of one component can affect many other components because of the complex network of interactions. If one component interacts directly with another (e.g., predators consuming prey), we call this a first-order effect. Insofar as interactions have been studied in ecosystem analysis, they have largely been of this direct or first-order type. However, the activities of one component may set off a chain of events that affects many components. We call these indirect, or higher-order, effects. For example, effects of predators on rates of primary productivity, via their regulation of herbivore grazing rates, would be referred to as a higher-order effect.

Higher-order effects in ecosystems are particularly evident in consumer populations. Decomposer arthropods in terrestrial ecosystems assimilate only a small fraction of the nutrients contained in leaf litter and have little direct effect on nutrient cycling [58]. However, as food passes through the gut, the material is finely ground, increasing the surface area available for microbial growth. In turn, the microbes assimilate a considerable fraction of the available nutrients and retard nutrient loss from the system [59,60]. Thus, although decomposer arthropods directly assimilate small quantities of nutrients, their feeding activities contribute significantly to nutrient cycling in the system [61].

Similar statements can be made about herbivorous arthropods in a variety of terrestrial ecosystems. Such herbivores typically consume a very small proportion of autotrophic net productivity, but may have profound indirect effects on a variety of ecosystem processes [62,63]. Mattson and Addy [64], in particular, hypothesized that arthropod defoliators, through a variety of indirect pathways, may act to regulate forest productivity over the long term. Studies of the fall cankerworm, *Alsophila pometaria,* a geometrid defoliator, have demonstrated the existence of such indirect effects [65]. Defoliation in upper elevation forest stands led indirectly to increased rates of nutrient (especially N) recycling. This may have resulted in a short-term fertilization of these stands existing on relatively infertile soils. Indeed, it was in indirect effect—that of cankerworms on soil nitrifying bacteria and hence on streamwater NO_3^- losses from the forests [66]—that led to the subsequent discovery by human observers of the defoliation, rather than by direct observation of an outbreak population of cankerworms.

Higher-order effects account for the "surprising" toxic effects we have seen in ecosystems. For example, using DDT to eliminate pest arthropods also results in reduced fecundity in birds of prey because the pesticide is concentrated as it passes through the food web as explained by Woodwell [67]. Similarly, microbial methylation of mercury sets off a chain of interactions that results in toxic concentrations in swordfish meat. Such examples make a strong case for the need to examine and understand the structure of interactions in ecosystems and to examine effects of toxicants on this interaction network. Indeed, Patten and Witkamp [30] suggested over a decade ago that "to understand ecosystems ultimately will be to understand networks."

Because such higher-order effects integrate the behavior of a variety of biological populations, together with associated physicochemical processes, they may provide us with especially sensitive points at which to measure effects of toxic substances on ecosystems. Higher-order effects, referred to below as integrative properties, may take many different forms. Bioaccumulation of chlorinated hydrocarbons within food chains integrates the dynamic behavior of a large segment of the ecosystem. Decomposition of complex molecules involves the activities of many faunal and microbial populations together with their organic and inorganic substrates. We must learn which higher-order effects will not only serve as sensitive indicators of significant environmental degradation but also can be practically included in testing protocols.

One result of considering the interaction network of ecosystems is an increased awareness of the importance of rate processes. In many higher-order effects, elimination of a component or a change in the number or types of interactions does not cause a problem; the problem arises from changes in the rates at which interaction processes occur. A clear example of this type of interaction can be seen in the role of fish in phosphorus dynamics of lakes [68]. Fish biomass contains a significant fraction of the phosphorus in the water column. It can be argued [69] that the slow, steady excretion of phosphorus by fish provides a reliable source for the maintenance of algal production during periods of low phosphorus availability. Thus, a toxic substance that does not measurably increase fish mortality but merely affects excretion rates of phosphorus in the fish could have far-reaching effects on productivity of the lake ecosystem. Likewise, toxic substances that inhibit microbial remineralization of phosphorus contained in fish bodies could have a profound effect.

Problems of Inferring Ecosystem Responses from Component Responses

A third relevant deduction from ecosystem theory is that the intact system will display measurable phenomena, which may be explained by, but can seldom be predicted from, knowledge of isolated components [70]. Hierarchical systems are structured vertically into levels by frequencies of observable system behaviors, and horizontally into subsystems by interaction strengths. High-frequency component behaviors determine intrasubsystem dynamics. However, interactions among subsystems at a given level occur at lower frequencies, and are independent of the microscopic structural detail within subsystems [22]. Thus, as already emphasized, component behaviors are in some sense smoothed, averaged or filtered as they move upward through the hierarchy, and may not be directly observable at some higher level of interest and measurement.

Nutrient cycling is an example of a macroscopic property, measurable on the total ecosystem level, but difficult to predict from component behavior. Reichle et al. [29] presented turnover times (i.e., standing crop divided by output rate) for nitrogen in components and a total forest ecosystem. Component turnover times ranged from 0.02 yr for decomposers to 109 yr for soil. However, the calculation for the total forest showed a turnover time of 1815 yr, the latter resulting from the recycling of nitrogen among system components. Waide et al. [71] discussed similar results for forest calcium cycles.

Once a configuration of interactions has been established, the ecosystem may show remarkable stability properties. In a theoretical study, O'Neill and Giddings [72] showed that the productivity of a community of phytoplankton populations was far less variable than the productivity of the individual populations. Total productivity was controlled by available nutrients. Elimination of a population had little effect on total productivity since it resulted in a competitive release of other populations in the system, which increased their productivity by utilization of freed nutrients. Similarly, total grassland productivity has been shown to be relatively constant from year to year despite major shifts in species composition and relative abundance [73,74]. Patten [56] has argued that these properties are the logical consequence of system evolution.

It also appears that expected correlations may exist at the ecosystem level even though they cannot be established at the subsystem or component level. For example, a number of experimental (e.g., Hurd et al. [75]) and theoretical (e.g., May [76]) studies have attempted to establish a relationship between system stability and complexity. These studies have focused on the community level of organization and have failed to establish the expected relationship. On the other hand, Van Voris et al. [49] tested the hypothesis on an intact excised soil system and found a positive correlation for ecosystem-level measures even though no correlation could be demonstrated for population or community measures. Although the tests were conducted on microcosms and not on an entire ecosystem, such as a forest, these authors argued that the intact soil cores retained much of the biotic and abiotic complexity and interrelationships which determine system stability under natural conditions.

Other examples of macroscopic system properties can be cited. Allen et al. [77] found distinct nodes of community organization in lake phytoplankton only after performing a principal-components ordination of species first-differences based on weekly samples for one year. These nodes or organizational states could not be inferred from abundance data on any single taxon or group of taxa. Similarly, Waide et al. [37] found definable regions of system behavior in a two-dimensional pH-DO phase space for a series of aquatic laboratory microcosms, as explained earlier. Again, neither the presence nor the characteristics of these regions could be inferred from abundances or presence-absence records of any biotic components of these systems. In a slightly different approach, Emanuel et al. [78] demonstrated changes in spectral properties of total forest biomass in simulations of forest succession in the southern Appalachians with and without the inclusion of American chestnut. However, specific differences in spectral characteristics of the two simulated time series could not be related to any properties of chestnut or any other species in the simulated forest community.

One ecosystem phenomenon of particular relevance to toxic effects is species replacement. When an ecosystem is affected, species most sensitive to the impact may be eliminated from the system. However, the impact on total system properties, such as total productivity or biomass, may be minimal, because competitive release simply frees other populations to grow and fill the vacancy. The result is that ecosystem properties are relatively constant in response to the impact, and this response is achieved by eliminating sensitive species.

The possibility for species changes or replacements to buffer changes in total system properties has been called "congeneric homotaxis" [17,21]. Frequently, there are a number of functionally similar and perhaps taxonomically related components in "parallel" or redundant positions in ecosystem interaction networks. If such species have different environmental tolerances, changes in total system behaviors may indeed be buffered by species replacements. The extent to which this mechanism plays a role in maintaining ecosystem homeostasis remains unknown. However, to the extent that it is an important control mechanism in ecosystems, it will further complicate any attempts to infer system properties from component properties.

The phenomenon of species replacement is important for two reasons. On one hand, it obscures our understanding of laboratory results. An effect on a single population may result in its replacement by other, less sensitive species, with no overall measurable changes in ecosystem dynamics. On the other hand, the elimination of a sensitive species by competitors may be an unacceptable result when viewed by human society. The obvious case is when the eliminated species happens to be *Homo sapiens*. More commonly, the species has a particular value to society as a game species, a rare or endangered species, or simply because of esthetic values associated with the organism. Elimination of the sensitive species may be "acceptable" to the ecosystem but unacceptable to society. For example, the loss of the American chestnut from the eastern deciduous forest did not destroy the forest, but it certainly affected the furniture industry.

Taken together, the discussions in this section of our chapter strengthen and illuminate the points raised earlier. The main thrust of the argument is simply that ecosystem-level effects of toxic substances must be measured at the system level. Such ecosystem responses can neither be inferred nor predicted from measurements on isolated components in the laboratory. In the next section, we suggest some approaches to meeting this problem.

RECOMMENDATIONS FOR TOXIC SUBSTANCES RESEARCH

Thus far in our presentation, our prognosis for predicting toxic chemical effects on ecosystems has been largely negative. We have argued that ecosystem responses will "surprise" us if our predictions are based solely on measurements of isolated components. Ultimately, a new perspective, based on a refined understanding of ecosystem dynamics, must be developed. Such an encompassing theory of the ecosystem, however, lies in the future. Current ecological theory can provide only broad, general guidelines for ecosystem-level test protocols. Nonetheless, we have attempted to derive some general recommendations from the foregoing discussions. These recommendations are presented sequentially, though they overlap broadly and each one builds on the previous one.

Focus on Properties of Intact Test System

We should utilize test systems that are excised as intact units from natural environments. Size is not the key variable here, nor is taxonomic composition. What is essential is that the eventual test system contains a high degree of structural organization based on shared evolutionary histories of biotic components, i.e., the test system must behave as a functionally intact system. Single-species test systems, or test systems assembled arbitrarily from stock cultures in the laboratory (so-called gnotobiotic systems [55]) will not suffice in this regard.

Such an intact test system should not be viewed as a homolog of the ecosystem. We do not expect its responses to be identical to those of natural ecosystems, but we do expect the test system to alert us to the potential occurrence of a higher-order effect, even if this effect is due to changes in component interactions that we did not anticipate. In many respects, we would hope that the intact test system would be more sensitive to a toxic effect than any natural ecosystem. The situation is directly analogous to painting high concentrations of a chemical onto the skin of a cancer-sensitive white mouse. The purpose is not to predict how many humans will die of skin cancer but to alert society to a potential effect so that appropriate regulatory actions can be taken. Therefore, we want the test to be conservative rather than exact, and we want it to reflect system-level, and not only component-level, properties.

Focus on Integrative Properties of Intact Test Systems

Not only must we employ functionally intact test systems, we must also choose measurements that reflect macroscopic, system-level properties as opposed to microscopic or component-level properties. Measured parameters should result from or reflect properties of couplings among biological, physical and chemical processes operating within test systems. Only integrative properties of the intact test system will provide the information needed.

Focus on Integrative Biogeochemical Properties of Intact Test Systems

We have argued that a biogeochemical perspective on system dynamics is an appropriate and useful view of ecosystems. Thus, we suggest that test protocols for toxic substance evaluations should focus on integrative biogeochemical properties. Variables such as total CO_2 efflux, amounts of some element lost via leaching or flowthrough, or system dynamics in a two-dimensional pH-pE (Eh, DO) phase space have proved useful in this regard.

We pointed out earlier that many types of measurements

could indicate higher-order effects in the system. For example, changes in the fecundity of top carnivores integrate many causal links in the system. However, many of these indicators may be impractical in microcosms and are best viewed as potential monitoring points in the field. Recent microcosm experimentation has shown that integrative biogeochemical variables are more precise and more sensitive at detecting system-level effects than any population- or community-level measure [34, 37,49].

Focus on Rate Processes as Integrative Biogeochemical Properties of Intact Test Systems

Among the biogeochemical properties of intact test systems that could be measured, we suggest that rate-dependent processes would be the most useful. Storage of elements in some components of test systems will be rather slow to change, and should not be a sensitive indicator of responses to toxic substance additions. In other components, storage may change rapidly but, because of species replacement, provide little information on system response. In contrast, rate processes (e.g., rates of nitrogen mineralization or substrate decomposition) should change rapidly in response to toxic substance perturbations and, in addition, should be relevant indicators of system stress.

Because of interactions among species in intact test systems (or in natural ecosystems), elimination of a single population or shifts in relative abundances of species may have little effect on system properties. Which specific populations are performing a given function (e.g., photosynthesis) is of less significance than the rate at which that function proceeds. For example, information that a toxic substance decreases the net cellulase activity of a mixed, natural culture better reflects an ecosystem effect than does the information that a single isolated species shows decreased activity. Such a test would be technically more difficult to conduct in the laboratory, and one would expect that it would require larger samples to establish significance. Demonstration of an effect on a rate process (or functional group performing this function) has more direct implications for ecosystem function than measurements of components.

A particularly sensitive test might involve the decomposition rate of a complex molecule that requires sequential processing by organisms. In an intact test system, one would expect that relatively small effects on organisms at any stage of the process would take on the role of a rate-limiting step, involving buildup of intermediate products and possible inhibition of the entire process. Thus, a toxic effect on the overall rate process could be detected if any of the organisms in the system were adversely affected. However, we emphasize again that such measurements on rate processes must be made in the context of an intact test system, and not on isolated components.

We believe that these four general recommendations are consistent with our previous arguments and that they provide a reasonable starting point for the development of ecosystem-level test procedures. Such procedures are not meant to supplant current toxicological procedures, but to supplement them. Current approaches remain valuable components of an overall program for the management of toxic substance discharges into natural environments. To reiterate our main point, however, population- or organism-directed testing protocols will not provide sufficient information for predicting fates and effects of toxic substances released into natural environments. Ultimately, toxic chemicals affect organisms or populations only insofar as they are functioning components of whole ecosystems and certainly not as isolated components. Furthermore, demonstration of potential toxic effects on ecosystems can only be made at the total system level, and not at the component level. This viewpoint does not deny the value of examining species-level toxic responses, or of attempting to determine possible dose rates to man through extrapolation and epidemiologic studies as explained by Reiches [79]. It does suggest, however, that such approaches represent only part of the total problem and do not allow us to understand behaviors of potentially toxic substances released into intact, natural environmental systems. The population and related approaches must be supplemented by appropriate ecosystem-level testing protocols if a complete program of management and regulation of toxic substance discharges is to be developed.

Much research remains to be done to develop further the theoretical rationale on which our arguments are based and its practical implementation. Taken together, our recommendations would require the use of laboratory microcosms excised as intact units from natural environments. Neither artificially assembled nor fragmented systems will suffice.

This is not to claim that measurements made on intact microcosms can be immediately and directly extrapolated to natural environments. Dangers involved in such an extrapolation are nearly as great as those involved in any extrapolation based on isolated components. Much remains to be done to work out practical details of microcosm testing and to elaborate constraints on applying laboratory results to field situations. We must understand much more completely than we do now how to establish correspondences between integrative properties of intact test systems and comparable properties of natural ecosystems. Moreover, we must gain additional experience in recognizing which of many possible integrative measures of the "state of the whole" ecosystem—total CO_2 efflux, element losses in leachate, system dynamics in pH-Eh phase space, rates of substrate decomposition, or even a variety of possible measures of bioaccumulation of toxic chemicals in food chains—will give us the most directly useful information for predicting the effects of a given potentially toxic substance on natural ecosystems. We currently see no alternative for examining potential effects of toxic substances at the ecosystem level of biological organization.

REFERENCES

1. Tansley, A. G. "The Use and Abuse of Vegetational Concepts and Terms," *Ecology* 16:284–307 (1935).

2. Lindeman, R. L. "The Trophic Dynamic Aspect of Ecology," *Ecology* 23:399–418 (1942).
3. Odum, E. P. *Fundamentals of Ecology,* 1st ed. (Philadelphia: W. B. Saunders Co., 1953).
4. Evans, F. C. "Ecosystem as the Basic Unit in Ecology," *Science* 123:1227–1228 (1956).
5. Overton, W. S. "Decomposability: A Unifying Concept?" in *Ecosystem Analysis and Prediction,* S. A. Levin, Ed. (Philadelphia; SIAM, 1975), pp. 297–298.
6. Whittaker, R. H., and G. M. Woodwell. "The Evolution of Natural Communities," in *Ecosystem Structure and Function,* J. A. Wiens, Ed. (Corvallis: Oregon State University Press, 1972), pp. 137–160.
7. Lane, P. A., G. H. Lauff and R. Levins. "The Feasibility of Using a Holistic Approach in Ecosystem Analysis," in *Ecosystem Analysis and Prediction,* S. A. Levin, Ed. (Philadelphia: SIAM, 1975), pp. 111–129.
8. Odum, H. T. "Macroscopic Minimodels of Man and Nature," in *Systems Analysis and Simulation in Ecology, Vol. IV,* B. C. Patten, Ed. (New York: Academic Press, Inc., 1976), pp. 249–280.
9. Salt, G. W. "A Comment on the Use of the Term Emergent Properties," *Am. Nat.* 113:145–148 (1979).
10. Schindler, J. E., J. B. Waide, M. C. Waldron, J. J. Hains, S. P. Schreiner, M. L. Freedman, S. L. Benz, D. P. Pettigrew, L. A. Schissel and P. J. Clarke. "A Microcosm Approach to the Study of Biogeochemical Systems. 1. Theoretical Rationale," in *Microcosms in Ecological Research*, J. P. Giesy, Ed. DOE Symposium Series (in press).
11. O'Neill, R. V., and D. E. Reichle. "Dimensions of Ecosystem Theory," in *Forests: Fresh Perspectives from Ecosystem Analysis,* R. W. Waring, Ed. (Corvallis: Oregon State University Press, in press).
12. Simon, H. A. "The Architecture of Complexity," *Proc. Am. Philos. Soc.* 106:467–482 (1962).
13. Conrad, M. "Patterns of Biological Control in Ecosystems," in *Systems Analysis and Simulation in Ecology, Vol. IV*, B. C. Patten, Ed. (New York: Academic Press, Inc., 1976), pp. 434–456.
14. Weinberg, G. *An Introduction to General Systems Thinking* (New York: John Wiley and Sons, 1975).

15. Allen, T. F. H., and T. B. Starr. *Hierarchy: Perspectives for Ecological Complexity* (Chicago, IL: University of Chicago Press, in press).
16. Waide, J. B., and J. R. Webster. "Engineering Systems Analysis: Applicability to Ecosystems," in *Systems Analysis and Simulation in Ecology, Vol. IV*, B. C. Patten, Ed. (New York: Academic Press, Inc., 1976), pp. 329–371.
17. Hill, J., IV, and S. L. Durham. "Input, Signals, and Control in Ecosystems," Proc. 1978 IEEE Int. Conf. on Acoustics, Speech, and Signal Processing, Tulsa, OK, April 1978, pp. 391–397.
18. Rosen, R. "Some Systems Theoretical Problems in Biology," in *The Relevance of General Systems Theory,* E. Laszlo, Ed. (New York: Braziller, 1972) pp. 43–66.
19. Blackburn, T. R. "Information and the Ecology of Scholars," *Science* 123:1141–1146 (1973).
20. Weiss, P. A. "The Basic Concept of Hierarchic Systems," in *Hierarchically Organized Systems in Theory and Practice,* P. A. Weiss, Ed. (New York: Hafner Publishing Co., 1971), pp. 1–44.
21. Hill, J., IV, and R. G. Wiegert. "Microcosms in Ecological Modeling," in *Microcosms in Ecological Research,* J. P. Giesy, Ed. DOE Symposium Series (in press).
22. Simon, H. A. "The Organization of Complex Systems," in *Hierarchy Theory: The Challenge of Complex Systems,* H. H. Pattee, Ed. (New York: Braziller, 1973), pp. 1–28.
23. Allen, T. F. H., and H. H. Iltis. "Overconnected Collapse to Higher Levels: Urban and Agricultural Origins, A Case Study," in *Systems Science and Science,* B. H. Banathy, Ed. (Louisville, KY: Society for General Systems Research, Systems Science Institute, 1980), pp. 96–103.
24. Patten, B. C., R. W. Bosserman, J. T. Finn and W. G. Gale. "Propagation of Cause in Ecosystems," in *Systems Analysis and Simulation in Ecology, Vol. IV,* B. C. Patten, Ed. (New York: Academic Press, Inc., 1976), pp. 457–579.
25. Webster, J. R. "Hierarchical Organization of Ecosystems," in *Theoretical Systems Ecology: Advances and Case Studies,* E. Halfon, Ed. (New York: Academic Press, Inc., 1979), pp. 119–129.
26. Hutchinson, G. E. "Circular Causal Systems in Ecology," *Ann. N.Y. Acad. Sci.* 50:221–246 (1948).

27. Morowitz, H. J. "Physical Background of Cycles in Biological Systems," *J. Theor. Biol.* 13:60–62 (1966).
28. Morowitz, H. J. "The Derivation of Ecological Relationships from Physical and Chemical Principles," *Proc. Nat. Acad. Sci., U.S.* 71:2335–2336 (1974).
29. Reichle, D. E., R. V. O'Neill and W. F. Harris, "Principles of Energy and Material Exchange in Ecosystems," in *Unifying Concepts in Ecology*, W. H. van Dobben and R. H. Lowe-McConnell, Eds. (The Hague: Dr. W. Junk, 1975), pp. 27–43.
30. Patten, B. C., and M. Witkamp. "Systems Analysis of ^{134}Cesium Kinetics in Terrestrial Microcosms," *Ecology* 48:813–824 (1967).
31. Woodwell, G. M. "Toxic Substances and Ecological Cycles," *Scientific Am.* 216(3):24–31 (1967).
32. Loucks, O. L. "Contaminants and Recycling in Relation to Biogeochemical Cycles," in *Challenging Biological Problems: Directions Toward Their Solution*, J. A. Behnke, Ed. (New York: Oxford University Press, 1972), pp. 297–312.
33. Van Hook, R. I., W. F. Harris and G. S. Henderson. "Cadmium, Lead, and Zinc Distributions and Cycling in a Mixed Deciduous Forest," *Ambio* 6:281–286 (1977).
34. O'Neill, R. V., B. S. Ausmus, D. R. Jackson, R. I. Van Hook, P. Van Voris, C. Washburne and A. P. Watson. "Monitoring Terrestrial Ecosystems by Analysis of Nutrient Export," *Water Air Soil Poll.* 8:271–277 (1977).
35. Baas-Becking, L. G. M., I. R. Kaplan and D. Moore. "Limits of the Natural Environment in Terms of pH and Oxidation-Reduction Potential," *J. Geol.* 68:243–284 (1960).
36. Stumm, W., and J. J. Morgan. *Aquatic Chemistry* (New York: Wiley Interscience, 1970).
37. Waide, J. B., J. E. Schindler, M. C. Waldron, J. J. Hains, S. P. Schreiner, M. L. Freedman, S. L Benz, D. P. Pettigrew, L. A. Schissel and P. J. Clarke. "A Microcosm Approach to the Study of Biogeochemical Systems. 2. Responses of Aquatic Laboratory Microcosms to Physical, Chemical, and Biological Perturbations," in *Microcosms in Ecological Research*, J. P. Giesy, Ed. DOE Symp. Series (in press).
38. Bormann, F. H., and G. E. Likens. "Nutrient Cycling," *Science* 155:424–429 (1967).
39. Likens, G. E., F. H. Bormann, N. M. Johnson and R. S. Pierce.

"The Calcium, Magnesium, Potassium, and Sodium Budgets in a Small Forested Ecosystem," *Ecology* 48:772–785 (1967).

40. Johnson, P. L., and W. T. Swank. "Studies of Cation Budgets in the Southern Appalachians on Four Experimental Watersheds with Contrasting Vegetation," *Ecology* 54:70–80 (1973).
41. Elwood, J. W., and G. S. Henderson. "Hydrologic and Chemical Budgets at Oak Ridge, Tennessee," in *Coupling of Land and Water Systems*, A. D. Hasler, Ed. (New York: Springer-Verlag, 1975), pp. 31–51.
42. Henderson, G. S., W. T. Swank, J. B. Waide and C. C. Grier. "Nutrient Budgets of Appalachian and Cascade Region Watersheds: A Comparison," *For. Sci.* 24:385–397 (1978).
43. Swank, W. T., and J. B. Waide. "Interpretation of Nutrient Cycling Research in a Management Context: Evaluating Potential Effects of Alternative Management Strategies on Site Productivity," in *Forests: Fresh Perspectives from Ecosystem Analysis*, R. W. Waring, Ed. (Corvallis: Oregon State University Press, 1980).
44. Likens, G. E., F. H. Bormann, N. M. Johnson, D. W. Fisher and R. S. Pierce. "Effects of Forest Cutting and Herbicide Treatment on Nutrient Budgets in the Hubbard Brook Watershed-Ecosystem," *Ecol. Monog.* 49:23–47 (1970).
45. Likens, G. E., F. H. Bormann, R. S. Pierce, J. S. Eaton and N. M. Johnson. *Biogeochemistry of a Forested Ecosystem* (New York: Springer-Verlag, 1977).
46. Swank, W. T., and J. E. Douglass. "Nutrient Flux in Undisturbed and Manipulated Forest Ecosystems in the Southern Appalachian Mountains," Publication No. 117, L'association Internationale des Sciences Hydrologiques. Symposium de Tokyo, December 1975.
47. Monk, C. D., D. A. Crossley, Jr., R. L. Todd, W. T. Swank, J. B. Waide and J. R. Webster. "An Overview of Nutrient Cycling Research at Coweeta Hydrologic Laboratory," in *Watershed Research in Eastern North America*, D. L. Correll, Ed. (Edgewater, MD: Smithsonian Institution, 1977).
48. Bormann, F. H., and G. E. Likens. *Pattern and Process in a Forested Ecosystem* (New York: Springer-Verlag, 1979).
49. Van Voris, P., R. V. O'Neill, W. R. Emanuel and H. H. Shugart, Jr. "Functional Complexity and Ecosystem Stability," *Ecology* 61:1352–1360 (1980).

50. Gardner, M. R., and W. R. Ashby. "Connectance of Large Dynamic (Cybernetic) Systems: Critical Values for Stability," *Nature* 228:784 (1970).
51. Pattee, H. H. "The Complementarity Principle in Biological and Social Structures," *J. Soc. Biol. Structures* 1:191–200 (1978).
52. DeAngelis, D. L. "Stability and Connectance in Food Web Models," *Ecology* 56:238–243 (1975).
53. Tregonning, K., and A. Roberts. "Complex Systems Which Evolve Toward Homeostasis," *Nature* 281:563–564 (1979).
54. Elton, C. S. *The Ecology of Invasions by Animals and Plants* (London: Chapman and Hall, 1958).
55. Taub, F. B. "Closed Ecological Systems," *Ann. Rev. Ecol. Syst.* 5:139–160 (1974).
56. Patten, B. C. "Ecosystem Linearization: An Evolutionary Design Problem," *Am. Nat.* 109:529–539 (1975).
57. Caswell, H., H. E. Koenig and Q. E. Ross. "An Introduction to Systems Science for Ecologists," in *Systems Analysis and Simulation in Ecology, Vol. III*, B. C. Patten, Ed. (New York: Academic Press, Inc., 1972), pp. 3–78.
58. Cornaby, B. W., C. S. Gist and D. A. Crossley, Jr. "Resource Partitioning in Leaf-Litter Faunas from Hardwood and Hardwood-Converted-to-Pine Forests," in *Mineral Cycling in Southeastern Ecosystems*, F. G. Howell, J. B. Gentry and M. H. Smith, Eds., ERDA Symp. Series, CONF-740513. (Springfield, VA: National Technical Information Service, 1975), pp. 588–597.
59. Ausmus, B. S., N. T. Edwards and M. Witkamp. "Microbial Immobilization of Carbon, Nitrogen, Phosphorus, and Potassium: Implications for Forest Ecosystem Processes," in *The Role of Terrestrial and Aquatic Organisms in Decomposition Processes*, J. M. Anderson and A. MacFadyen, Eds. (London: Blackwell, 1976), pp. 397–416.
60. Witkamp, M., and B. S. Ausmus. "Processes in Decomposition and Nutrient Transfer in Forest Systems," in *The Role of Terrestrial and Aquatic Organisms in Decomposition Processes*, J. M. Anderson and A. MacFadyen, Eds. (London: Blackwell, 1976), pp. 375–396.
61. Witkamp, M. "Soils as Components of Ecosystems," *Ann. Rev. Ecol. Syst.* 2:85–110 (1971).
62. Chew, R. M. "Consumers as Regulators of Ecosystems: An Alternative to Energetics," *Ohio J. Sci.* 74:359–370 (1974).

63. Owen, D. G., and R. G. Wiegert. "Do Consumers Maximize Plant Fitness?" *Oikos* 27:488–492 (1976).
64. Mattson, W. J., and N. D. Addy. "Phytophagous Insects as Regulators of Ecosystems," *Science* 190:515–522 (1975).
65. Crossley, D. A., Jr., W. T. Swank, R. L. Todd and J. B. Waide. Unpublished data.
66. Swank, W. T., and J. E. Douglass. A Comparison of Nutrient Budgets for Undisturbed and Manipulated Hardwood Forest Ecosystems in the Mountains of North Carolina," in *Watershed Research in Eastern North America,* D. L. Correll, Ed. (Edgewater, MD: Smithsonian Institution, 1977), pp. 343–364.
67. Woodwell, G. M. "Toxic Substances: Clear Science, Foggy Politics," Chapter 2, this volume.
68. Kitchell, J. F., J. R. Koonce and P. S. Tennis. "Phosphorus Flux Through Fishes," *Verh. Int. Verein. Limnol.* 19:2478–2484 (1975).
69. Kitchell, J. F., R. V. O'Neill, D. Webb, G. W. Gallepp, S. M. Bartell, J. F. Koonce and B. S. Ausmus. "Consumer Regulation of Nutrient Cycling," *BioScience* 29:28–34 (1979).
70. Ghent, A. W. "On the Anatomy of Explanation, Part 3. Sequential and Hierarchical Description." *Biosphere* 34:11–23 (1963).
71. Waide, J. B., J. E. Krebs, S. P. Clarkson and E. M. Setzler. "A Linear Systems Analysis of the Calcium Cycle in a Forested Watershed Ecosystem," in *Progress in Theoretical Biology, Vol. 3,* R. Rosen and F. M. Snell, Eds. (New York: Academic Press, Inc., 1974), pp. 261–345.
72. O'Neill, R. V., and J. M. Giddings. "Population Interactions and Ecosystem Function," in *Systems Analysis of Ecosystems,* G. S. Innis and R. V. O'Neill, Eds. (Fairland, MD: International Cooperative Publishing House, 1979), pp. 103–123.
73. Bement, R. E. "Dynamics of Standing Dead Vegetation on the Short Grass Plains," in *The Grassland Ecosystem: A Preliminary Synthesis,* R. L. Dix and R. G. Beidleman, Eds. (Fort Collins: Colorado State University, 1969), pp. 221–224.
74. McNaughton, S. J. "Diversity and Stability of Ecological Communities: A Comment on the Role of Empiricism in Ecology," *Am. Nat.* 111:515–525 (1977).
75. Hurd, L. E., M. V. Mellinger, L. L. Wolf and S. J. McNaughton. "Stability and Diversity at Three Trophic Levels in Terrestrial Successional Ecosystems," *Science* 173:1134–1136 (1971).

76. May, R. M. *Stability and Complexity in Model Ecosystems* (Princeton, NJ: Princeton University Press, 1973).
77. Allen, T. F. H., S. M. Bartell and J. F. Koonce. "Multiple Stable Configurations in Ordinations of Phytoplankton Community Change Rates," *Ecology* 58:1076–1084 (1977).
78. Emanuel, W. R., H. H. Shugart, Jr. and D. C. West. "Spectral Analysis and Forest Dynamics: The Effects of Perturbations on Long-Term Dynamics," in *Time Series and Ecological Processes*, H. H. Shugart, Jr., Ed. (Philadelphia: SIAM, 1978), pp. 193–207.
79. Reiches, N. A. "Environmental Carcinogens: The Human Perspective," Chapter 3, this volume.

5

Paradigms in Multiple Toxicity

Perry D. Anderson
Department of Biological Sciences
Concordia University
Sir George Williams Campus
Montreal, Quebec
Canada

Chemical pollutants seldom exist alone in the environment. Rather, they coexist in varying kinds and quantities. Consequently, exposure to a single substance is the exception, and multiple exposure to two or more chemicals is the rule.

Physiological interactions may occur between chemical contaminants during their respective but overlapping fates within an organism. Interactions may arise at sites of toxic action in target tissue and/or during kinetic processes of absorption, deposition, distribution, excretion and metabolic transformation. Such interplay may lead to adverse effects that would not have occurred, at least not to the same degree and/or kind, with a single chemical contaminant.

Toxicity response patterns attributable to physiological interactions among two or more substances are classified as multiple toxicity. The term is not limited to a particular type of endpoint response (e.g., lethal, sublethal, carcinogenic, mutagenic or teratogenic), or restricted to a particular mode of exposure (e.g., acute, chronic or intermittent). It also includes the toxicity that may arise between chemical contaminants whose respective routes of uptake into an organism are either similar or different.

Obviously, multiple toxicity is one of contemporary society's most insidious as well as ubiquitous Medusas. Central to the issue of how to tame or defend against this Medusa is the question whether environmental quality criteria for individual substances provide adequate protection. Except for mixtures associated with overt emissions from point sources, environmental quality objectives are based on toxicological impact assessments of single substances. Thus, there are no regulations for mixtures that arise from overlap in the environmental vectors of their constituents. Since ad hoc mixtures are found not only in habitats closely associated with man's agricultural, industrial and municipal centers, but in wilderness regions, it would be an immense if not impossible task, at least in the foreseeable future, to identify and characterize their potential for multiple toxicity.

Environmental quality objectives for safeguarding the health of organisms, including man, against single substance toxicity are usually set at or around empirically defined thresholds of tolerance [1]. As seen in Figure 1, such thresholds are best clarified when based on dose-response data that are accumulated from numerous toxicological studies. Forms of multiple toxicity that are deemed hazardous arise from coexisting pollutants whose respective levels are within the permissible, i.e., "safe," range. The following described patterns of multiple toxicity meet these high risk criteria.

ADDITIVE PARADIGMS IN MULTIPLE TOXICITY

Concentration Addition

The simplest pattern of multiple toxicity is called concentration addition [2]: it arises when two or more subthreshold sub-

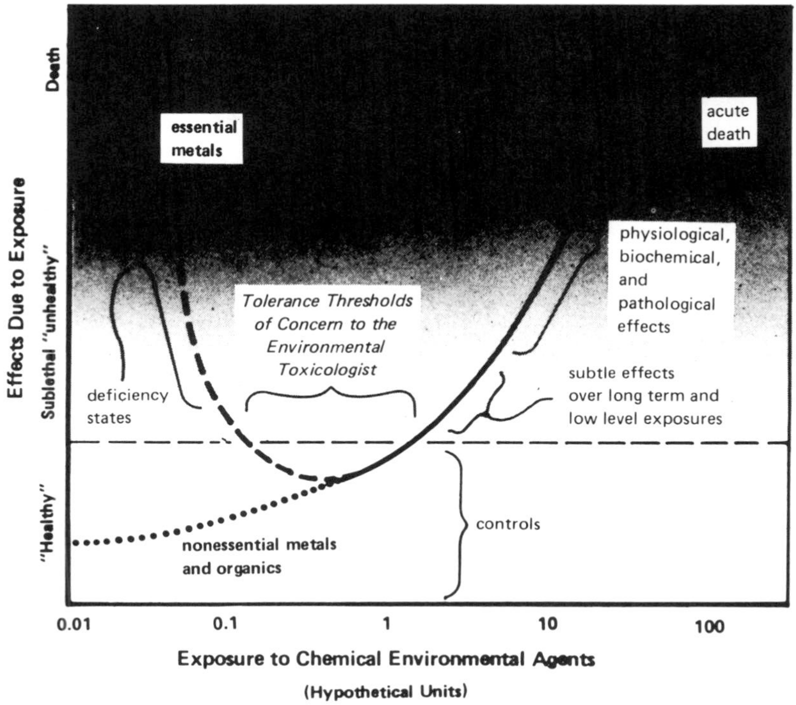

Figure 1. Overview of significant concerns to environmental toxicologists for establishing relationships between exposure to a chemical environmental agent and the effects of exposure.

stances contribute to a threshold or greater effect in proportion to their respective relative potencies. The substances are assumed to act similarly. The concept of concentration addition is demonstrated for a binary mixture of similarly acting compounds in Figure 2. The "Before Exposure" part of Figure 2 shows the total available binding sites within a hypothetical target tissue. A specific proportion of these sites, say 40%, must be occupied if an adverse response is to be elicited. No harm is incurred if a lesser proportion of the sites are bound, because organisms are able to adapt physiologically to a low-grade disturbance in tissue function and maintain normal organ activity. Thus neither the concentration of A nor that of B in the hypo-

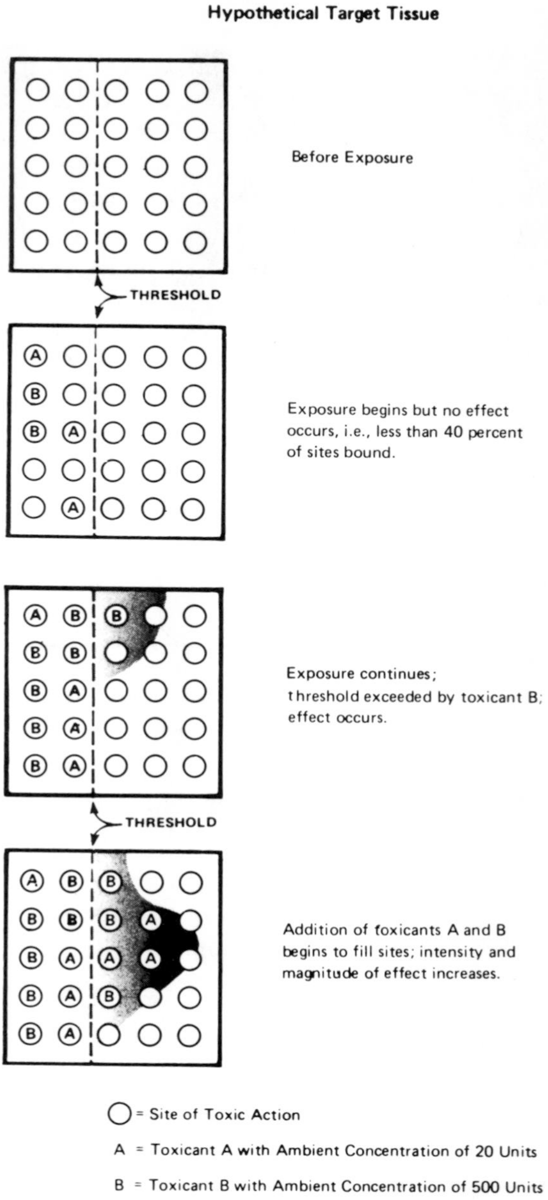

Figure 2. Sequence for addition of similarly acting constituents to produce a threshold and greater effect.

thetical mixture is sufficient alone to bind the necessary proportion of sites to evoke an effect. However, when A and B act together, the combined bound sites exceed the threshold number of 10, and a response follows. Note that the potency of toxicants which act similarly can and usually do differ, which means that it takes a greater exposure to one toxicant (B), than it does to the other (A) to occupy the same number of sites. With consideration for differences in potency, the magnitude of the effect elicited by similarly acting constituents can be predicted by the following formula:

$$E\left(\frac{1}{n}\sum_{j=1}^{n} C_{sj}\right) = E(C_{s1}) = \ldots = E(CSn) \qquad (1)$$

where E = a common effect
$Cs_1, \ldots Cs_n$ = concentrations of similarly acting constituents that produce a particular magnitude of effect [3]

If the magnitude of E for the mixture is significantly greater than predicted by Equation 1, then the joint response is called supraadditive, and it is assumed that some form of synergistic interaction has occurred [3]. If the magnitude of E is significantly less than expected, the toxicity pattern is called infra-additive, and some form of antagonism is assumed. Equation 1 provides a means of empirically testing whether water quality objectives for single agents provide protection against the multiple toxicity expressed in hazardous form of concentration addition.

Response Addition

The above forms of concentration addition, including the subsets of infra and supra, can be distinguished from another type of joint additive effect where the constituent metals in a mixture act quite independently (i.e., at different target sites) and yet contribute to a common whole-organism response, e.g., death or reduction in growth rate. In theory, each independently acting constituent can be expected to contribute to a common response order if its respective concentration is equal to, or greater than, the threshold computed from its own dose-response curve. The magnitude of response elicited by a mixture whose constituents act independently at concentrations at or above their respective thresholds for discrete solutions is calculated as follows:

$$P_m = 1 - (1 - P_1)(1 - P_2) \ldots (1 - P_n) \qquad (2)$$

where P_m = proportion of individuals responding to a mixture
$P_1, P_2, \ldots P_n$ = proportion of individuals responding to pure solutions of each constituent at concentrations $X_1, X_2, \ldots X_n$, respectively [4]

Equation 2 assumes that there are neither kinetic nor dynamic interactions between the independently acting constituents and that there is no correlation between tolerances of the test organisms to each of the poisons. Anderson and Weber [2] called this toxicity pattern response addition. It is often mistaken for concentration addition and deemed likewise to be a hazardous form of multiple toxicity. Rather, the respective mechanisms are quite unlike. The basic assumption of independent action, unlike that of similar action, supports the rationale that safe or no effect levels can be established for an individual toxicant without concern for multiple toxicity due to interactions with other environmental contaminants.

Nonadditive Forms of Multiple Toxicity

Nonadditive forms of multiple toxicity would be of particular concern when the combined effects are different from, as well as more toxic than those predicted from a knowledge of the toxicity of the respective constituents. Anderson and d'Apollonia [3] attempted to categorize various nonadditive forms of multiple toxicity, including potentiation, sensitization and permissive. Another form would involve interactions within mixtures containing essential metals that would lead to a deficiency state as indicated in Figure 1. Methodologies for quantifying these nonadditive forms of multiple toxicity are generally lacking. However, some qualitative characteristics of these nonadditive joint toxicities are cited later in this chapter.

GENERAL AND SPECIFIC TOXICITY-BASED REGULATIONS

Metal mixtures, particularly those that contain cadmium (Cd), copper (Cu), mercury (Hg) and zinc (Zn), are of concern, because they are among the most biologically available elements in aquatic ecosystems and because their respective toxicity thresholds are often close to the ambient levels that prevail through natural geological forces [5]. Ambient levels may approximate toxicologically effective concentrations where man's activities promote a high rate of metal mobilization, particularly via acid precipitation [6,7]. Also, there is evidence suggesting a link between combinations of metal contaminants and perturbations in aquatic populations [8–10].

There is a great complexity among the toxicities reported for individual metals. One might reasonably expect that this complexity would be compounded for the toxicity of metal mixtures, and that the adverse effects of the latter would not simply be characterized by the mechanism of similar action, as seen in a concentration addition response pattern. Yet there have been suggestions that metals have a common capacity to combine with organic molecules at ligands and thereby denature both structural and enzymatic proteins [11]. Furthermore, there have been frequent reports of additivity for metal mixtures in aquatic environments [12]. Thus, some authorities have assumed that there is a general additive mechanism for metals, and have proposed a toxic unit formula for waters contaminated with mixtures of metal pollutants [13]. Application of the formula means exceedingly stringent emission controls and profound economic consequences.

One of our objectives was to investigate whether such a

singular, all-encompassing standard for metal mixtures was justifiable. Our methodology was adapted from Bliss [14], who presented a rationale for testing the hypothesis of similar action as assumed in Equation 1. He proposed that the normally distributed variance in tolerance of a test population exposed to different poisons (that act alike) should be the same, and that this fact would be displayed empirically by parallelism between dose-response lines for similarly acting constituents and their mixtures. Given that the respective dose-response lines for individual pollutants have a similar slope, the response curve for their mixture may be predicted using the formula:

$$Y_m = a + b \log [\pi X_1 + (1 - \pi) p X_2] \qquad (3)$$

where Y_m = probit of response for mixture
X_1 = concentration of constituent 1
X_2 = concentration of constituent 2
π = proportional of the total amount combined
p = relative potency factor between constituents 1 and 2
b = common slope to discrete response curves of constituents 1 and 2 [3]

The assumption of similar action is tested by comparing the predicted line or curve to a curve fitted to the observed responses to the mixture [2]. If the empirically derived points lie to the right of the predicted curve, the combined effect is called infraadditive. If the empirical curve lies significantly to the left of the predicted curve, the combined effect is called supraadditive.

EXAMPLES

Examples of multiple toxicity patterns arising from selected binary mixtures of heavy metals are now provided. There is particular emphasis on cadmium-zinc, cadmium-mercury, nickel-vanadium and mixtures of copper with cadmium, nickel and zinc.

Effects of Cadmium and Zinc Mixtures

In a series of studies in which zebra fish (*Brachydanio rerio*) were used as the test organism and in which binary mixtures of certain heavy metals were examined, only Cd-Zn mixtures strictly obeyed the model of concentration addition (see Equation 3). Figure 3 shows that the line fitted to the dose-response data for lethal mixtures of Cd-Zn falls within the 95% confidence limits for the predicted concentration addition function. The same pattern was observed for embryonic, larval and juvenile stages in the life cycle. However, the four life-cycle stages did differ in their tolerance to Cd, Zn and their mixtures. Larvae were the most sensitive, followed by juveniles, adults and eggs, in that order.

Further tests showed that the difference in lethal tolerances between each stage could be correlated with the change in weight-specific metabolic rate with age [15]. This correlation suggests that Cd and Zn have a similar mode of action within a particular life stage and that this action is common to all stages examined. Furthermore, the action would appear to be nonspecific, being common to the fertilized egg and the developing embryo in which tissue and organ targets are absent or not fully developed. Our hypothesis is that Cd and Zn, through protein denaturation, affect the boundary membranes involved in respiration. There is apparent justification for applying to natural water bodies, subject to Cd and Zn contamination, a toxic unit standard [12], that provides a safety factor against a concentration addition pattern of toxicity.

Effects of Cadmium and Mercury Mixtures

Concentration addition appeared to represent the multiple toxicity pattern of Cd and Hg mixtures, but only when the exposure period was approximately four days. The pattern dif-

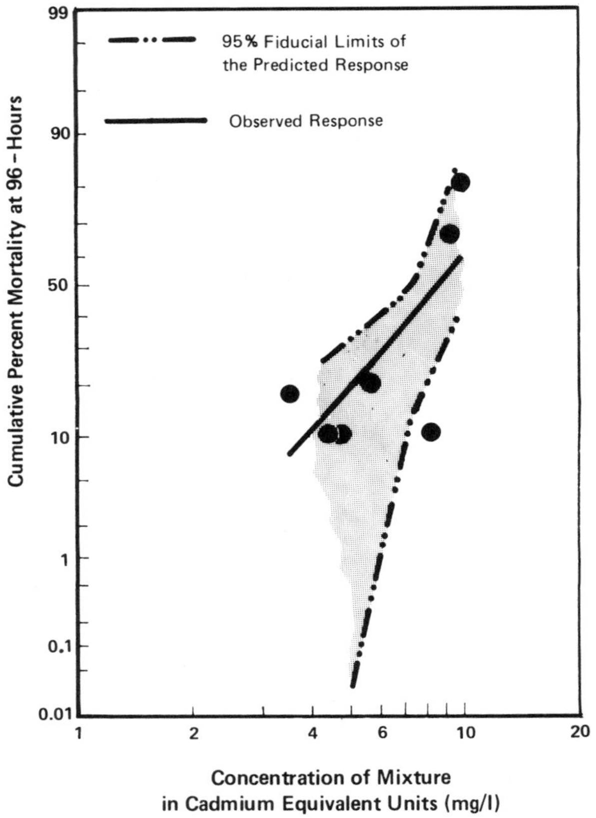

Figure 3. The response of the adult zebra fish exposed to mixtures of Cd and Zn conforms to the model of concentration addition.

fered with lesser or greater exposure. There may be a time-related interaction between Cd and Hg coincidental with the magnitude of effect predicted by the empirical model for similarly acting substances at 96 hours of exposure. Furthermore, the toxicity observed with Cd-Hg mixtures would seem to be a property of Hg alone, because the behavioral signs of toxicity for the Cd-Hg exposed organisms were characteristic of individuals exposed to pure solutions of Hg. Related studies suggested the subthreshold presence of cadmium in Cd-Hg mix-

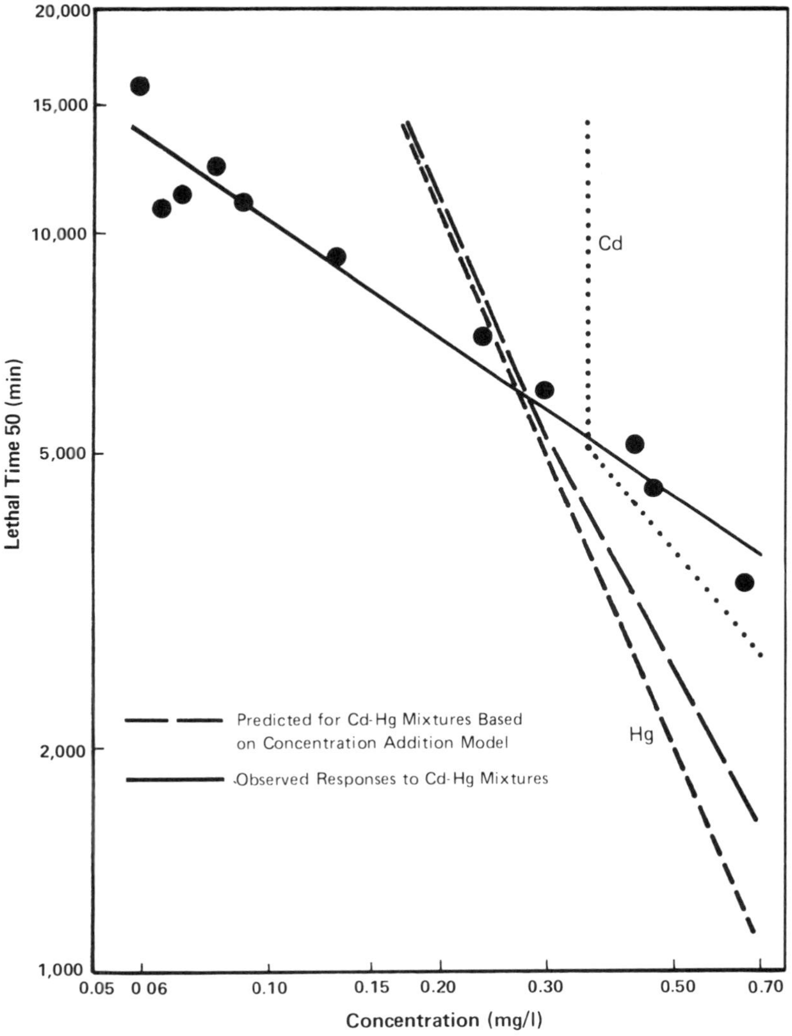

Figure 4. Toxicity curves for zebra fish exposed to pure solutions of Cd or Hg and their mixtures.

tures were altering the availability of Hg to critical target tissue through time [16].

Some aspects of the effects of this apparent interaction are illustrated in the distribution of curves [time to 50% mortality (LT50)] shown in Figure 4. Note that the Cd curve displays an incipient lethal level (ILL) within 96-hours (5760 min). At concentrations below the ILL, at least 50% of the test organisms were able to adapt and avoid the directly lethal effects of Cd. This capacity suggests that Cd is a noncumulative agent. However, the toxicity curve for pure solutions of Hg was linear over the 14 days of the study. Such a distribution (1) infers that the test organisms did not adapt within this time period and (b) is consistent with the observation of other researchers that Hg is a cumulative toxicant. The broken line indicates the expected distribution of points for Cd-Hg mixtures on the assumption that the constituents are concentration additive (see Lloyd [17] and Herbert and Shurben [18] for methodology that predicts time-related responses according to the concentration addition model). The observed responses to Cd-Hg mixtures (filled circles) would appear to follow a linear distribution that crosses the predicted line at about four days. Prior to this time the observed responses are categorically infraadditive, while subsequently they are supraadditive. By the 14th day, the mixture was approximately threefold more toxic than was predicted by the concentration model. Contrary to the antagonistic role of Cd on Hg toxicity in mammals [19], Cd significantly enhances the lethal toxicity of this cumulative metal (Hg), at apparent sublethal levels and for long-term coexposure. In contrast to its applicability for Cd-Zn mixtures, the Cd-Hg toxic unit standard [12] for protection of fishes in water would not appear to be adequate.

Effects of Nickel and Vanadium

Another incident of infraaddition was recorded in toxicity studies of lethal concentrations of Ni and V, in which trout (*Salmo gairdneri*) were the test organism. The magnitude of infraaddition was found to increase progressively as the propor-

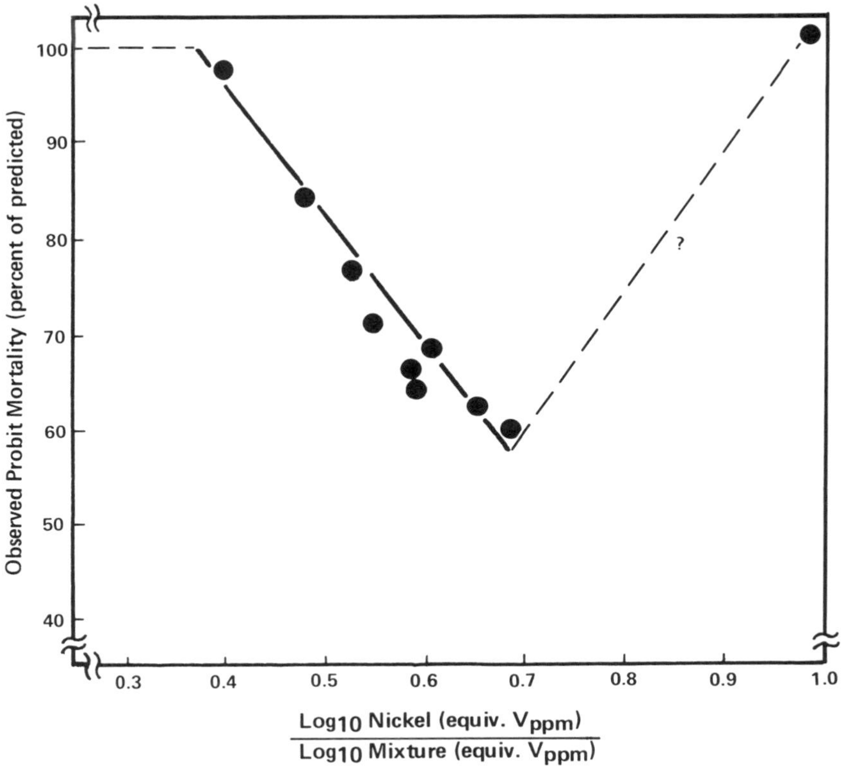

Figure 5. Decrease in lethal potency of Ni-V mixtures from that predicted for concentration addition. Antagonism increases as the ratio of Ni to total concentration increases.

tion of Ni to V increased within mixtures. In mixtures where the ratio of Ni to V was approximately equal, the response was 40% less than that predicted on the basis of the concentration addition model [20] (Figure 5). Observations were made at 96 hours of exposure. Other studies in our laboratory on binary mixtures that contain Ni, but not V, have not displayed an infraadditive pattern. Thus, V is suspected to be the antagonistic agent. It would be interesting to compare the biological role of V to that of selenium (Se)—a known antagonist—to certain heavy metal toxicity, e.g., Hg [21].

Effects of Copper and Cadmium, Nickel and Zinc Mixtures

Several binary mixtures Cu-Cd, Cu-Ni and Cu-Zn, in which Cu was a common constituent were examined in accordance with Equation 3. All were significantly supraadditive, even though it had been assumed that constituents were acting similarly because of the parallelism of their individual dose-response lines. Additional research explored the possibility that kinetic interactions may be a factor in creating the unexpectedly greater than additive response. Uptake rates into gill tissue—a suspected critical target site for these heavy metals [22]—were measured for concentration of each element when present alone in pure solutions and when present together with one of the other metals—Cd, Ni and Zn. Figure 6 for Cd-Cu mixtures shows that Cu uptake rates were significantly enhanced by association with Cd. Conversely, the uptake of the Cd constituent was increased in the presence of Cu (Figure 7). Supraadditive effects for certain mixtures containing Cu occur also for aquatic weeds [23] and algae [24].

The consistent appearance of this multiple toxicity pattern among diverse aquatic organisms suggests a common mechanism of interaction between the boundary membrane transport pathways for Cu and those of certain other metals, e.g., Cd, Ni and Zn. As in the case of Cd-Hg mixtures, the toxic unit formula for safeguarding aquatic biota from the toxicity of multicontaminants would appear to be inadequate for the above listed copper-containing mixtures.

MORE QUALIFICATIONS

There are at least two factors that temper the interpretation of biological effects to chemical mixtures. One, the quality of the water, whether soft or hard, makes a difference. Second, the relationships of binary mixtures for lethal situation may change for sublethal thresholds. Each of these will be discussed.

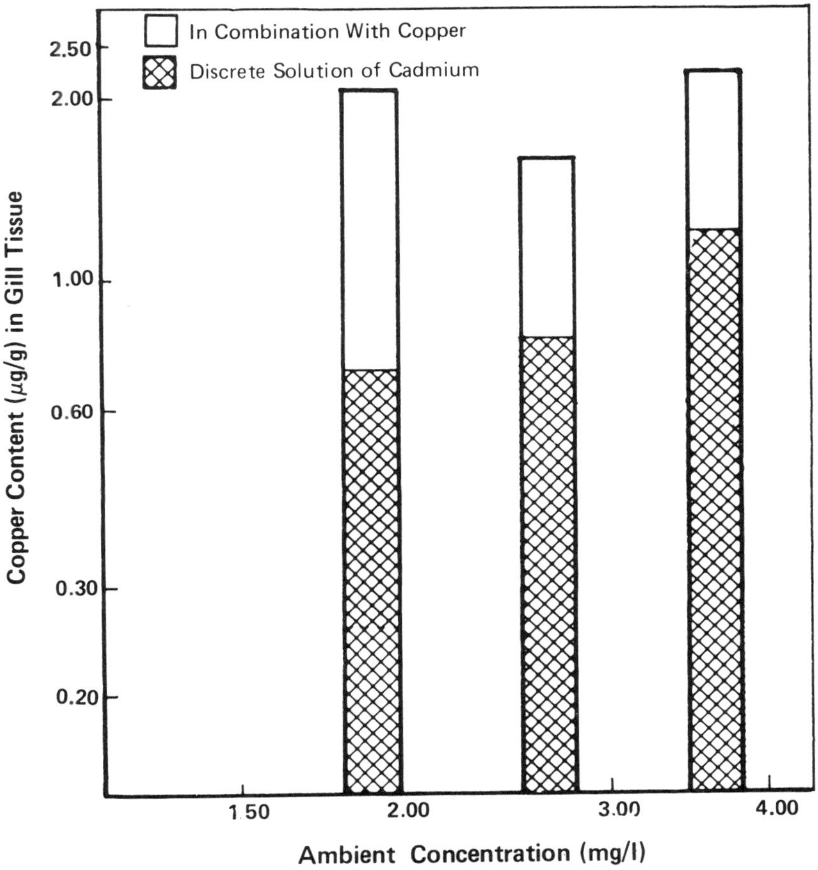

Figure 6. Copper content in fish gill tissue following a 20-hr exposure to Cu alone and in combination with Cd. The binary combinations contained (left) 1.880, (center) 2.626 and (right) 3.565 mg/l Cd.

Water Softness and Hardness

In water quality management, the setting of heavy metal standards may take into consideration certain biotic and environmental variables that are known to influence the toxicity of metals. The environmental factor, water hardness as Ca^{2+}

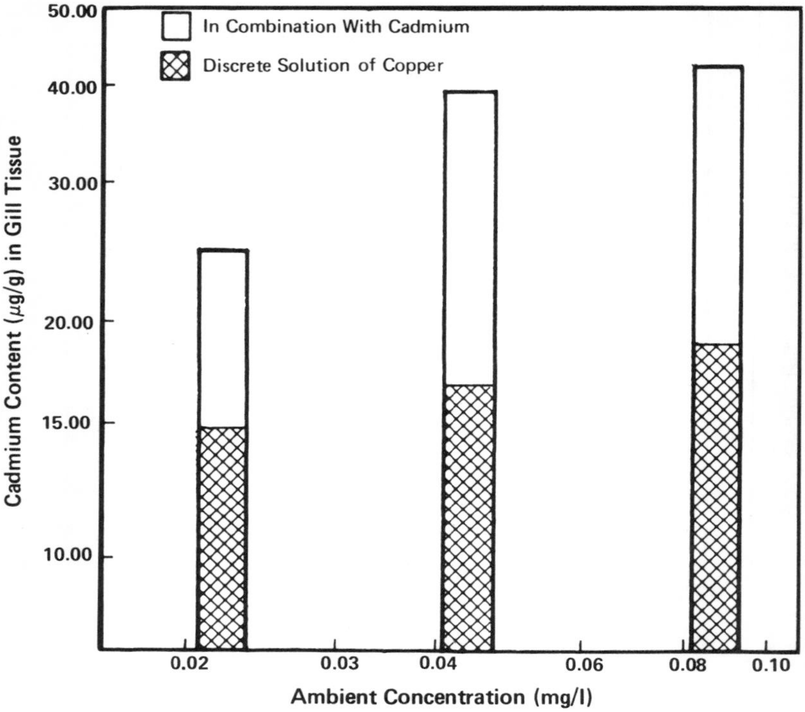

Figure 7. Cadmium content in fish gill tissue following a 20-hr exposure to Cd alone and in combination with Cu. The binary combinations contained (left) 0.021, (center) 0.043 and (right) 0.088 mg/l Cu.

and Mg^{2+} ions, is a significant modifier of metal toxicity. Formulations have been developed [25,26] that estimate the decrease in toxicity of certain heavy metals with increasing hardness. Alkalinity and pH were held constant. The LC50 concentrations are in Cu equivalent units adjusted for the modifying effect of sodium ions. When applied to water quality objectives, these formulations permit higher ambient levels of metals in hard waters than they do in soft waters. A study in our laboratory suggests that the apparent reduction in the toxicity of a metal in increasing hard water may not be so great when other metals are present concurrently. Figure 8 shows that mixtures of Cu and Zn are approximately concen-

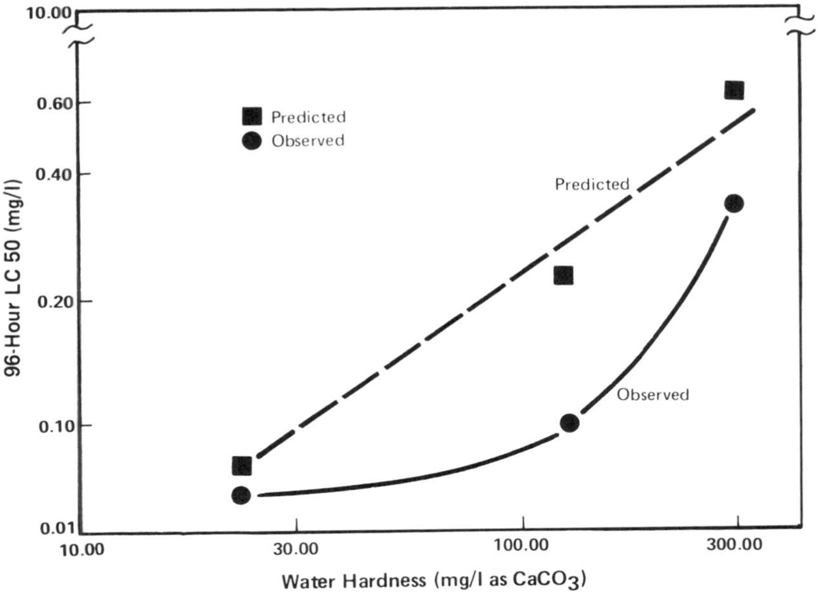

Figure 8. The toxicity of Cu-Zn mixtures over a range of water hardness shows how this environmental parameter modifies toxicity.

tration additive in soft waters. However, whereas the toxicity of both metals as discrete agents decreased dramatically with increasing water hardness (see example for Cu), the toxicity of mixtures of Cu and Zn did not decrease to the same degree. Although concentration additive in soft waters, Cu-Zn mixtures are significantly supraadditive in harder (120–300 mg/l as $CaCO_3$) waters. This pattern suggests that any leniency that would be granted in water quality objectives for single heavy metals in hard waters may not be applicable if other heavy metals are present concurrently.

Applicability to Sublethal Thresholds

All previous studies have dealt with lethal thresholds, and it may be asked whether thresholds to sublethal events (see

Figure 1) are subject to the same patterns of multiple toxicity. We have examined this question for mixtures of Cu and Ni. As a single lethal agent, Cu is about 170-fold more potent than Ni. Because their respective dose-lethality curves were parallel, their mixtures were tested for concentration addition. The observed pattern for their mixtures was supraadditive, the enhancement in lethal toxicity being sixfold greater than expected [27]. An examination of the sublethal effects of pure solutions of Cu or Ni on egg production by test populations of zebra fish showed the same difference in relative potency as in lethal studies, i.e., Cu was 170-fold more toxic as a sublethal agent than Ni. When sublethal studies were conducted with Cu-Ni mixtures, on the assumption of concentration addition, total inhibition of egg production occurred. The actual responses predicted are indicated by the Ni equivalent response

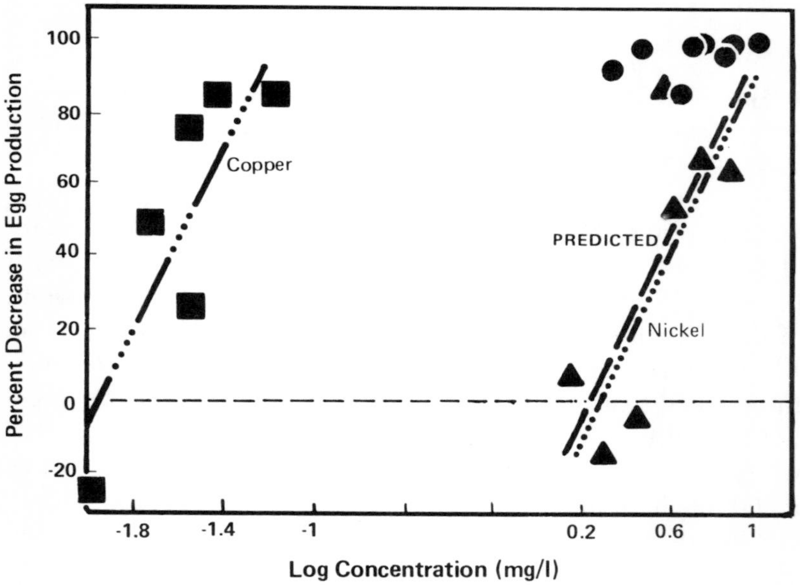

Figure 9. Mixtures of Cu and Ni inhibit egg production (closed circles). Cu and Ni alone also inhibited egg production. The predicted dose-response curve for the mixture is similar to the one for Ni alone.

line in Figure 9. Concentrations are represented in Ni equivalents. Note that low levels of either Cu or Ni are actually stimulatory. At least for Cu and Ni mixtures, the pattern of supraaddition exists for both lethal and sublethal-reproductive response regimes. Thus, the toxic unit standard [12] is applicable to neither lethal nor sublethal thresholds where Cu-Ni mixtures are concerned.

OTHER DIMENSIONS OF MULTIPLE TOXICITY

Other studies confirm that many heavy metal mixtures are highly toxic to a variety of organisms other than fish. This can be true even when each chemical constituent is individually at a subthreshold level. Physiological interactions that lead to detrimental effects commonly occur among metals that are jointly present in a target organism. This likelihood negates the use of single contaminant water quality criteria that will protect organisms other than fish in habitats subject to multimetal pollution. Studies by Wong et al. [28] support this conclusion. These researchers found that the growth of algae was severely inhibited in culture media contaminated by ten heavy metals, each of which was present at its maximal permissible concentration (i.e., at an apparently harmless level). Even reducing the maximal permissible concentration of each metal by a factor of 10, in accordance with the toxic unit criteria for additivity, did not prevent a significant reduction in the growth rate of algae. The result suggests supraadditive interactions among the metals. Several of the metals used in this study were observed in our own studies to cause supraaddition as binary mixtures.

Hazardous forms of multiple toxicity are not limited to heavy metal mixtures. Physiological interactions also occur between subthreshold levels of organic pesticides to create adverse responses. Supraaddition was observed in 20%, while concentration addition characterized 55% of the mixtures of 47 binary mixtures of pesticides, including carbamates, hydrocarbons and organophosphates [29,30]. Furthermore, a recent survey [31] concluded that pesticide mixtures were generally more than additive to aquatic organisms.

Another potential threat of multiple toxicity exists in mixtures where certain inorganics are combined with organics. Maines and Kappas [32] have shown—at least for mammals—that various metals, but particularly Cd and cobalt (Co) inhibit mixed-function oxidase (MFO) systems. Many naturally produced and foreign organics are metabolized by MFO systems. Jernelöv et al. [33] suggested that suppression or induction of these microsomal enzyme systems could explain many of the metabolic interactions arising from the combined presence of different chemicals and often leading to potentiated effects. MFO-mediated interactions have been shown for piperonyl butoxide and several carbamates, organophosphates and organochloride pesticides. Common industrial organics such as carbon tetrachloride, chloroform, Kepone and mirex are thought to interact—at least in part—through MFO systems [34]. MFO systems are known to exist in fish [35] and therefore similar hazardous interactions may occur in this group of organisms.

Complex mechanisms of physiological interactions are now suspected in the ontology of certain tumors. Bingham et al. [36] proposed that some forms of cancer are caused by DNA-binding agents whose efficacy is mediated by promoter substances, called cocarcinogens, encountered either concurrently or subsequently. Cocarcinogens, although a requirement in this form of tumorigenesis, are not believed to be carcinogenic in themselves. Tumors have been found in fish inhabiting multipolluted waters [37]. Given that chemical carcinogenesis has been demonstrated empirically in fish [38], a combined action of pollutant chemicals to cause tumors cannot be ruled out.

Another dimension of multiple toxicity is the measurement of complex mixtures by exposing organisms to the entire chemical mixture, not a component of the mixture. Duke and Merrill [39] discuss this.

SUMMARY

Our results show that the toxicities of heavy metal mixtures cannot adequately be described by a singular paradigm, such as additivity. Some combinations appear to be additive (Cd-Zn), many are supraadditive (Cu-Ni, Cu-Cd, Cd-Hg and Cu-Zn) and one is apparently infraadditive (Ni-V).

The evidence also suggests that the toxicity pattern of a metal mixture may be subject to modification by various factors. For example, the environmental factor, water hardness, did not depress the lethal potency of Cd-Zn mixtures to the same degree that it did Cd or Zn alone. Even the type of the exposure may be a significant determinant in the response to mixtures. For example, the relative proportion of constituents within Ni-V mixtures and the duration of exposure to combinations of Cd and Hg influenced the toxic response. The relative potency of a mixture may change with the type of response observed in organic-inorganic mixtures [40], although in our studies of the inorganic mixture, Cu-Ni, both lethal and sublethal responses were supraadditive.

There is little evidence to suggest that biological factors such as age and species affect the pattern of multiple toxicity of heavy metals. Tolerance may vary, however, between species and within life cycles. No one seems to have examined whether the route of uptake significantly modifies multiple toxicity.

There was an apparent parallelism between the dose-response curves for several of the heavy metals examined singly in our laboratory. However, the use of parallelism as a reflection of similar action [14] and thereby as a criterion for additivity did not prove reliable. In general, the probability of estimating the toxicity of mixtures from an evaluation of whole organism responses to each constituent alone is low.

Hazardous forms of multiple toxicity would appear to be common to many metal mixtures. It is difficult to predict the magnitude and type of response that interacting metals evoke.

There is a need to understand the mechanisms of physiological interactions between metals with the objective of providing information that would permit the derivation of effective water quality criteria. Such a task requires suitable toxicity testing rather than trying to predict without data. Furthermore, this research should not be restricted to metal mixtures alone, but should include studies of mechanisms that apparently potentiate the toxicity of inorganic-organic and purely organic mixtures. Research should also be devoted to environmental carcinogenesis, a portion of which may be the consequence of insidious mechanisms of multiple toxicity. Then, the general approach to standards development can gradually be replaced by relatively specific standards for specific effluents.

ACKNOWLEDGMENTS

Much of the empirical data reported herein was part of the MSc research of my graduate students: P. Spear, L. Weinstein, H. Horovitch, B. Gallimore and A. Hewitt. S. d'Apollonia also contributed to these projects. Financial support was contributed mainly by the Water Subcommittee of the National Research Council, Contract #031-1042. Funds were also provided by the Department of Education, Quebec and by the Alberta Oil Sands Environmental Research Program.

REFERENCES

1. Warren, C. E. *Biology and Water Pollution Control.* (Philadelphia: W. B. Saunders Co., 1971).
2. Anderson, P. D., and L. J. Weber. "The Toxicity to Aquatic Populations of Mixtures Containing Certain Heavy Metals," in *Proc. International Conference on Heavy Metals in the Environment, Vol. 2.* (Toronto, Ontario, 1975), pp. 933–953.
3. Anderson, P. D., and S. d'Apollonia. "Aquatic Animals," in *Principles of Ecotoxicology,* Scope 12. G. C. Butler, Ed. (New York: John Wiley & Sons, 1978), pp. 187–221.

4. Finney, D. E. *Probit Analyses,* 3rd ed. (London: Cambridge University Press, 1971).
5. Wood, J. M., H. J. Segall, W. P. Ridley, A. Chen, W. Chudyk and J. S. Thayer. "Metabolic Cycles for Toxic Elements in the Environment," in *Proc. International Conference on Heavy Metals in the Environment, Vol. 1.* (Toronto, Ontario, 1975), pp. 49–68.
6. Waldichuk, M. "Some Biological Concerns in Heavy Metals Pollution," in *Pollution and Physiology of Marine Organisms* (New York: Academic Press, Inc., 1974), pp. 1–57.
7. Lewis, W. M. Jr., and M. C. Grant. "Changes in the Output of Ions from a Watershed as a Result of the Acidification of Precipitation," *Ecology* 60(6):1093–1097 (1979).
8. Van Loon, J. C., and R. J. Beamish. "Heavy Metal Contamination by Atmospheric Fallout of Several Flin Flon Area Lakes and the Relation to Fish Populations," *J. Fish Res. Board Can.* 34:899–906 (1977).
9. Whitby, L. M., P. M. Stokes, T. C. Hutchinson and G. Myslik. *Can. Mineral.* 14:47–57 (1976).
10. McFarlane, G. A., and W. G. Franzin. "Elevated Heavy Metals: A Stress on a Population of White Suckers, *Catastomus commersoni,* in Hamell Lake, Saskatchewan," *J. Fish Res. Board Can.* 35:963–970 (1978).
11. Passow, H., A. Rothstein and T. W. Clarkson. "The General Pharmacology of the Heavy Metals," *Pharmacol. Rev.* 13: 185–224 (1971).
12. Sprague, J. B. "Review Paper Measurement of Pollutant Toxicity to Fish. 2. Utilizing and Applying Bioassay Results," *Water Res.* 4:3–32 (1970).
13. Seba, D. B. "Toxicity Index for Permits," in *Proceedings of the Symposium on Structure-Activity Correlations in Studies of Toxicity and Bioaccumulation with Aquatic Organisms,* G. D. Veith, and D. E. Konasewich, Eds. Windsor, Ontario: I.J.C. Res. Adv. Board, 1975), pp. 199–259.
14. Bliss, C. I. "The Toxicity of Poisons Applied Jointly," *Ann. Appl. Biol.* 36:385–615 (1939).
15. Gallimore, B., and P. D. Anderson. "Life Cycle Patterns in Lethal Tolerance of Fish Exposed to Cadmium, Zinc and Their Mixtures. *Pharmacologist* 21(3):250 (1979).
16. Hewitt, L. A. "Dose and Time Related Response Patterns in

Test Populations of *Brachydanio rerio* Exposed to Copper, Cadmium and Mercury in Pure Solutions and in Binary Mixtures," MSc thesis, Concordia University, Montreal (1980).

17. Lloyd, R. "The Toxicity of Mixtures of Zinc and Copper Sulphates to Rainbow Trout, *Salmo gairdneri,* Richardson," *Ann. Appl. Biol.* 49:535–538 (1961).
18. Herbert, D. W. M., and D. G. Shurben. "The Toxicity to Fish of Mixtures of Poisons. I. Salts of Ammonia and Zinc," *Ann. Appl. Biol.* 53:33 (1964).
19. Schnell, R. C. "Cadmium-Induced Alternation in Drug Action," *Fed. Proc.* 37(1):28–34 (1978).
20. Anderson, P. D., P. Spear, S. d'Apollonia, S. Perry, J. de Luca and J. Dick. "The Multiple Toxicity of Vanadium, Nickel and Phenol to Fish," AOSERP Report 79, prepared for the Alberta Oil Sands Environmental Research Program (1979).
21. Burk, R. F., K. A. Foster, P. M. Greenfield and J. P. Kiper. "Binding of Simultaneously Administered Inorganic Selenium and Mercury to a Rat Plasma Protein," *Proc. Soc. Exp. Biol. Med.* 145:782–785 (1974).
22. Anderson, P. D., and P. A. Spear. "Copper Pharmacokinetics in Fish Gills—I. Kinetics in Pumpkin Seed Sunfish, *Lepomis gibbosus* of Different Body Sizes," *Water Res.* 14:1101–1105 (1980).
23. Hutchinson, T. C., and H. Czyrska. "Heavy Metal Toxicity and Synergism to Floating Aquatic Weeds," *Verh. Int. Verein. Limnol.* 19:2102–2111 (1975).
24. Hutchinson, T. C. Personal communication.
25. Andrew, R. W. "Toxicity Relationships to Copper Forms in Natural Waters," in *Toxicity to Biota of Metal Forms in Natural Waters,* R. W. Andrew, P. V. Hodson and D. E. Konasewich, Eds. (Windsor, Ontario: Great Lakes Research Advisory Board, International Joint Commission, 1976), pp. 127–143.
26. Spear, P. A., and R. C. Pierce. "Copper in the Aquatic Environment: Chemistry, Distribution, and Toxicology," NRCC 16454. (Ottawa, Ontario: Environmental Secretariat, 1979), p. 227.
27. Weinstein, N., and P. D. Anderson. "Lethal and Sublethal Toxicities of Copper-Nickel Mixtures to the Zebra Fish, *Brachydanio rerio,*" Can. Fish. Mar. Serv. Tech. Rep. No. 818 (1978), pp. 153–167.
28. Wong, P. T. S., Y. K. Chan and P. L. Lexon. "Toxicity of Mixtures of Metals on Freshwater Algae," *J. Fish. Res. Board Can.* 35:479–481 (1978).

29. Marking, L. L., and W. L. Mauck. "Toxicity of Paired Mixtures of Candidate Forest Insecticides to Rainbow Trout," *Bull. Environ. Contam. Toxicol.* 13(5):518–523 (1975).
30. Macek, K. J. "Acute Toxicity of Pesticide Mixtures to Bluegills," *Bull. Environ. Contam. Toxicol.* 14(6):648–651 (1975).
31. J. S. Alabaster, Ed. *Combined Effects on Freshwater Fish and Other Life of Mixtures of Toxicants in Water*, European Inland Fisheries Advisory Commission, Subcommission III (UK), FAO (1980), Report 37, p. 49.
32. Maines, M. D., and A. Kappas. "Metals as Regulators of Heme Metabolism," *Science* 198:1215–1221 (1977).
33. Jernelöv, A., K. Beijer and L. Soderlund. "General Aspects of Toxicology," in *Principles of Ecotoxicology*, Scope 12. G. C. Butler, Ed. (New York: John Wiley & Sons, 1978), pp. 151–168.
34. Hewitt, W. R., M. G. Coté and G. L. Plaa. "Acute Alteration of Chloroform-Induced Hepato- and Nephrotoxicity by Mirex and Kepone," *Fed. Proc.* 34:402 (1978).
34. Elcombe, C. R., and J. J. Lech. "Induction and Characterization of Hemoprotein(s) P-450 and Monooxygenation in Rainbow Trout (*Salmo gairdneri*)," *Toxicol. Appl. Pharmacol.* 49: 437–450 (1979).
36. Bingham, E., R. W. Niemeier and J. B. Reid. "Multiple Factors in Carcinogenesis," *Ann. N.Y. Acad. Sci.* 271:14–21 (1976).
37. Brown, E. R., J. S. Hazdra, L. Keith, I. Greenspan, J. B. G. Kwapinski and P. Beamer. "Frequency of Fish Tumors Found in a Polluted Watershed as Compared to Nonpolluted Canadian Waters," *Cancer Res.* 33:189–198 (1973).
38. Wales, J. H., and R. O. Sinnhuber. "Hepatomas Induced by Aflatoxin in the Sockeye Salmon (*Oncorhznchus nerka*)," *J. Nat. Cancer Inst.* 48:1529–1530 (1972).
39. Duke, K. M. and R. G. Merrill, Jr. "Development of New Biological Test Protocols," Chapter 6, this volume.
40. Broderius, S. J., and L. L. Smith. "Lethal and Sub-lethal Effects of Binary Mixtures of Cyanide and Hexavalent Chromium, Zinc, or Ammonia to the Fathead Minnow (*Pimephales promilas*) and Rainbow Trout (*Salmo gairdneri*)," *J. Fish. Res. Board Can.* 36:164–172 (1979).

6

Development of New Bioassay Protocols

Kenneth M. Duke
Health and Environmental Sciences Section
Battelle Columbus Laboratories
Columbus, Ohio
Raymond G. Merrill, Jr.
Industrial Environmental Research Laboratory
United States Environmental Protection Agency
Research Triangle Park, North Carolina

The wastes from industrial and energy conversion processes constitute a major source of exposure for humans and other organisms. Large quantities of such wastes are released annually to the environment in gaseous, liquid and solid forms. Exposure by organisms is by direct inhalation, ingestion or absorption of the waste through food web transfer. Health and ecological effects from such exposure range from stimulation of growth in nutrient-limited organisms to irreversible changes, including death.

Recognition of the problems associated with the environmental release of wastes has resulted in environmental legislation, including the Clean Air and Clean Water Acts and their various amendments, and the Resource Conservation and Recovery Act (RCRA). This legislation enables federal and state governments to set and enforce emission standards which would control and manage the amounts of various pollutants released. To effectively implement these laws it is necessary to (1) identify which constituents of waste streams are actually hazardous or toxic, (2) determine acceptable release limits, and (3) have available efficient, cost-effective control technology capable of meeting the established limits. The U.S. Environmental Protection Agency (EPA) has developed an environmental assessment program for obtaining such information about waste streams.

The EPA environmental assessment sampling and analysis program, developed at their Industrial Environmental Research Laboratory at Research Triangle Park, NC, has four components [1,2]:

1. evaluation of biological, chemical and physical characteristics of all waste streams;
2. identification of probable health and ecological effects of streams;
3. ranking of waste streams according to relative potential hazard; and
4. identification of appropriate pollution control technology programs.

The ultimate objective of this program is the development of a control technology program using both process engineering changes and addition of specific control devices to reduce emissions to the prescribed limits. The development of such a program requires a thorough understanding of the properties of the emission streams needing control, which in turn requires some procedure to identify the nature, extent and magnitude of the problem stream. Physical, chemical and biological studies are necessary to obtain the required understanding of the waste streams. Physical data (e.g., temperature, pressure or flowrates) are useful in determining rates of emission and in selecting appropriate procedures for control. Chemical information will identify the known pollutants as well as other constituents in a stream and provide data on concentrations which will aid in directing control activities. Biological data will permit an overall assessment of a stream's toxicity, including that arising from known pollutants, compounds not yet identified as hazardous, and synergistic or antagonistic interactions among the various constituents.

Two approaches are possible for obtaining the necessary data to design a control technology program. The first is the direct approach. Each waste stream is subjected to a thorough sampling and biological, chemical and physical testing protocol. A relatively high degree of confidence is associated with the results of this approach. It is also expensive and time-consuming. A second approach, developed by the EPA [2] is called the phased approach. Studies are organized into three phases designed to reduce costs without compromising the accuracy or reliability of the resultant data. The initial phase (Level 1) acts as a screen using "grab" samples and simple tests (e.g., determination of chemical classes, static acute bioassays) to measure cost-effectively the basic chemical, physical and biological characteristics of each waste stream. Results are qualitative or, at best, semiquantitative and are used to rank the streams in order of relative hazard. The second phase (Level 2) involves more detailed sampling and analysis starting with the higher priority streams and working through the lesser priority streams as funds permit. Chemical tests will resolve the waste stream into its constituent compounds; biological tests will confirm and/or elaborate on level 1 results, providing more accurate data about the potential hazards. Finally, level 3 monitors specific pollutants of interest and the effects of various operating conditions on the streams. Chronic bioassays may be included in level 3, but levels 1 and 2 depend primarily on acute bioassays. The procedures for implementing the phased approach are under continual development and refinement. As of 1980, levels 1 and 2 sampling procedures and chemical analyses, and level 1 biological procedures were well along in development and refinement. Level 3 for sampling and chemical analyses and levels 2 and 3 for biological analyses were still under development. Because the phased approach is so reasonable, important and timely, the focus of the remainder of this chapter will be on the level 1 biological testing protocol and its implementation in selected EPA environmental assessment programs.

LEVEL 1 BIOLOGICAL TESTING PROTOCOL

Development of the current (1980) level 1 biological testing protocol began in 1976 when EPA formed its Bioassay Subcommittee composed of selected health and ecological effects scientists from several EPA research and development laboratories. During this four-year period, the subcommittee has applied the principles of the phased approach to the development of level 1 biological analysis procedures [3], field tested the procedures via pilot studies, and revised and refined the procedures to solve problems revealed by the pilot studies and subsequent implementation. The following describes the procedures for this important protocol.

Level 1 Sample Collection and Preparation

Level 1 begins with the proper collection, transport and preparation of samples of the waste streams from a facility [4]. Care must be taken to get the best possible samples within the time and cost constraints imposed at level 1. These constraints dictate "grab" type samples taken at a single point in time during the operating cycle of the facility. All waste streams—gaseous, liquid and solid—must be sampled.

Sample Collection and Transport

The large-volume gaseous sample needed for one of the biological tests—the plant stress ethylene—is collected using a sampling probe inserted into an appropriate port in the stack. If the stream is under positive pressure, the sampling bags, made of Tedlar® (a relatively inert material), are filled with the aid of a polypropylene regulating valve in the Teflon® sampling line to prevent bursting the bag. Negative pressure streams are sampled by placing an evacuated Tedlar sample bag connected to the sample line into a 55-gal fiberboard drum and evacuating the drum. As the negative pressure in the drum exceeds that of the waste stream, the bag will fill. These procedures are designed to prevent the gas stream from contacting anything but inert materials, thus reducing the possibility of contamination and preserving sample integrity to the best possible degree. A second, much smaller, sample of gaseous streams is taken using a source assessment sampling system (SASS), which removes suspended particles of various sizes using a series of cyclones and a filter and organics using a sorbent column [4]. The particulates and solvent samples are transported in amber glass, while large volume gas samples are shipped in the Tedlar bags protected by 55-gal fiberboard drums.

Liquid samples are collected in glass or linear polyethylene containers using a tap in the waste stream or a dipper or some other means for withdrawing the sample. Large volumes are required for the aquatic tests. Samples are cooled to 4°C for transport and storage prior to testing.

Simple grab samples using manual or auger (boring) collection techniques are suitable for solids. These samples are placed in amber glass or linear polyethylene containers for shipping and storage.

While the objective of level 1 sampling is to obtain representative waste stream samples, no attempt is made to collect them under either isokinetic or isothermal conditions nor to account for temporal variations in the process and its wastes.

Sample Preparation

There is a need for many of the samples to undergo preparation to make them acceptable for use in the biological tests (see Figure 1). For example, particulates and solids used in the health effects test needed to be respirable size (5 μm or less). Consequently, samples need to be sized to ensure the use of 5 μm or smaller particles. If the quantity of small particles is insufficient, grinding may be required. The extraction of organics from the sorbent columns used to filter gaseous or liquid samples usually involves the use of methylene chloride, which is incompatible with the health effects tests. Solvent exchange using dimethyl sulfoxide (DMSO) to replace the methylene chloride is required. Solid samples to be tested via aquatic tests must undergo leachate preparation. A modified ASTM-A (American Society for Testing and Materials) aqueous leach procedure is used to generate sufficient quantities of leachate for aquatic testing. Other sample preparation procedures include filtering larger particles (greater than 2mm) from liquid samples and concentrating some dilute aqueous samples for use in the Ames and rodent tests to raise the concentration of constituents above each test's threshold limit.

Thus, all level 1 sample preparations are done under carefully controlled conditions and using specific equipment and materials to prevent contamination of the sample.

LEVEL 1 BIOLOGICAL TESTS

Several criteria were used in the development of the level 1 protocol. Paramount, of course, was the need to assess the health and ecological effects of a wide diversity of waste streams. Toxicity and mutagenicity (presumed carcinogenicity) are the key effects to be detected by the health tests. The use of both in vitro and in vivo test systems was also a criterion for the selection of health effects tests. Measurement of toxicity was the key to the ecological test selection. Also, the inclusion of different kinds of organisms (e.g., microorganisms, plants and animals) and different habitats (e.g., aquatic, terrestrial and marine) was deemed important. Other criteria included (1) a short testing period, no longer than two or three weeks, that resulted in acute as opposed to chronic or life-cycle tests and (2) low cost for each test; each was generally to cost less than $1500. Finally, each test incorporated into the protocol

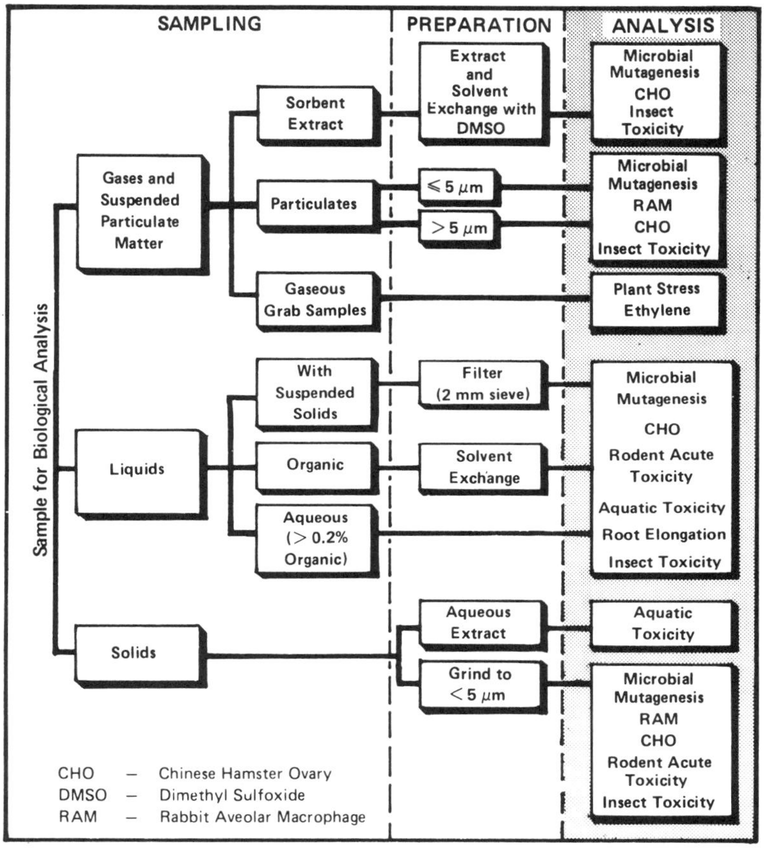

Figure 1. The level 1 biological testing protocol consists of a variety of organisms representing a variety of habitats. Organisms are exposed to chemical mixtures.

was to be acceptable by the scientific community as valid, reliable and accurate with a sufficiently large existing data base to facilitate the interpretation of the results. The application of these criteria resulted in the selection of the biological tests shown in Table I. The application of these tests to specific kinds of environmental samples is shown in Figure 1 [3].

All of the level 1 biological tests follow the same basic procedure. Test organisms are separated into groups, and each

Table I. 1980 EPA Level 1 Recommended Health and Ecological Effects Tests

Health Effects	Ecological Effects
In vitro	
Ames microbial mutagenesis assay (Ames)	Aquatic[a]
	Algal toxicity test
Rabbit alveolar macrophage cytotoxicity assay (RAM)	Static acute invertebrate toxicity test
Chinese hamster ovary cell clonal toxicity assay (CHO)	Static acute fish toxicity test
	Terrestrial
In vivo	Plant stress ethylene test
Rodent acute toxicity test	Seed germination–root elongation test
	Insect toxicity test

[a]Both marine and freshwater species are used in the aquatic effects tests.

group is exposed to one concentration of the test sample. Concentrations are fixed for each test and range from the maximum applicable doses down to some small fraction of that dose (e.g., 0.1 or 1%). There are usually five or six concentrations per test plus the negative control (zero concentration). Several tests use the results of this type of exposure as a rangefinding test and follow up with a definitive test with four or five concentrations bracketing the projected concentration (based on the rangefinding results) which produces the prescribed test endpoint. All tests have acute short-term exposures and use lethality or a similar measure (e.g., reduced growth) as the endpoint.

Health Effects Tests

The purpose of the health effects tests is to obtain data that will permit predictions of the potential hazard of complex waste streams to humans. The in vitro tests are designed to measure mutagenicity and toxicity at the cellular level. For example, the Ames microbial mutagenesis assay employs a mutant strain of the bacterium, *Salmonella,* as the test species and measures sample-induced mutation back to the normal form [5–7]. Cytotoxicity is also measured during this test, and results are expressed as the concentration at which the number of bacterial colonies is half that of the negative control (LC_{50}). While not a mammalian cell assay, it has been observed that many chemicals found to be carcinogenic in long-term mammalian studies also elicit a positive Ames test result. Thus, a positive Ames test identifies the potential of the sample being carcinogenic to mammals, including humans.

Cytotoxicity tests use mammalian cells to measure the toxic effects of samples [8–11]. The RAM test is used on respirable particulates (5 μm or less) and solid samples ground to respirable size. The CHO assay is used on liquid and solid samples. Both cell types are phagocytic and will ingest small particles. centration at which the responses of the experimental cultures centration at which the response of the experimental cultures reach one-half that of the controls. Cell viability and ATP content are the typical responses or endpoints measured. These results are an indicator of the response which might be observed in mammalian tissues of whole animals exposed to the sample. However, the in vitro nature of these tests greatly tempers any quantitative extrapolation to the whole animal.

Recognizing the difficulties in such extrapolations, an in vivo animal rodent acute toxicity test was included as part of level 1 [12–14]. Laboratory mice are used as the test organisms. Test endpoints are lethality (LD_{50}) plus pharmacological toxicity observations for nonlethal responses.

Ecological Effects Tests

The ecological effects tests are included in level 1 to provide an indication of the effects complex waste streams could have on organisms other than man. Species living in both aquatic and terrestrial habitats are tested since wastes are released to both the water and land. Finally, organisms are used that represent different levels of complexity ranging from unicells to higher plants and vertebrates to determine effects of wastes on different kinds of metabolic pathways and physiologies.

The aquatic ecological effect tests use algal, invertebrate and vertebrate species from both freshwater and marine habitats in acute toxicity testing. *Skeletonema costatum* and *Selenastrum capricornutum* are the marine and freshwater algal species, respectively. Invertebrates include *Daphnia magna* for freshwater testing and *Mysidopsis bahia* for marine testing. The marine vertebrate is the juvenile sheepshead minnow, *Cyprinodon variegatus,* while the fathead minnow, *Pimephales promelas,* is the freshwater vertebrate of choice. The vertebrates and invertebrates are tested using static acute procedures [15, 16]. The LC_{50} is the endpoint for these tests. The algal test protocol monitors changes in growth rate as a measure of toxicity [17]. The test endpoint is the effective concentration of sample which reduces algal growth by 50% as compared to the controls (EC_{50}). A second endpoint, the concentration at which increased growth is observed (stimulation concentration, SC), is determined for samples which may stimulate algal growth by supplying additional quantities of a normally limiting nutrient.

The level 1 terrestrial tests and test protocols had not as of 1980 been made final. The originally proposed tests [3] were unable to meet all the level 1 criteria as revealed by the pilot studies conducted in 1977. There have been subsequent modifications and refinements to the terrestrial protocol, but the ability of the resultant protocol to meet level 1 needs has not been fully validated. The current protocol includes the plant stress ethylene, seed germination–root elongation, and insect toxicity tests [18]. The plant stress ethylene test uses increased ethylene production as its endpoint [3,19]. Many plants show changes in ethylene production in response to stress; this response can be very sensitive and more replicable than the more traditional visual damage parameters often used to measure plant stress. The bushbean (*Phaseolus vulgaris*) is the test species. The test is used primarily on gaseous samples. Modifications to the basic test protocol are underway.

The seed germination–root elongation test, a recent addition to the level 1 ecological effects protocol, measures the effect of liquid or leachates from solid wastes on the germination and growth of several species of plants [20,21]. Test endpoints include percent germination and root length and results are expressed as the effective concentration required to reduce either measure by 50% (EC_{50}). The procedures for this test were still being refined and validated as of late 1980.

The final terrestrial test is the insect toxicity bioassay using *Drosophila melanogaster*. This test, also a new addition to level 1, is still to be validated. It involves acute exposures to various concentrations of liquid or solid samples with the test endpoint being lethality expressed as LD_{50}.

Quality Control and Quality Assurance (QC/QA)

The EPA is in the process of implementing QC/QA procedures to ensure the reliability and replicability of the level 1 biological test results. Such procedures range from proper labeling of a new sample through the use of a standard format for reporting test results. Furthermore, controls will be required for each test, a minimum being the negative (zero concentration) type of control with the addition of positive controls and solvent controls for certain tests. Standard test protocols must be written and available to the technical staff performing the bioassay. Any deviations from these standard protocols must be authorized by the EPA Project Officer and fully reported. Instruments must be calibrated regularly. There will be a strong emphasis on written documentation, and reports will be reviewed objectively for accuracy and consistency. Eventually all level 1 bioassays will be subject to the EPA Good Laboratory Practice (GLP) regulations.

Data Reporting and Interpretation

There are three levels of data reporting. The first is the raw test data. The actual laboratory data sheets are required showing data entries just as they were recorded in the lab. The second level reporting provides the toxicity or mutagenicity calculation (e.g., LC_{50} and EC_{50}) for each test. The final level is a nonnumerical evaluation of toxicity useful for a quick overview of the bioassay results and, when used with the level 1 chemistry data, permits the relative ranking of waste streams. Conversions from LC/EC data to these relative rankings are given (Table II).

APPLICATION OF LEVEL 1 BIOASSAYS TO FLUIDIZED BED COMBUSTION SAMPLES

One major EPA environmental assessment program where the phased approach has been applied, was a study of the fluidized bed combustion (FBC) process. FBC offers a feasible alternative for burning high-sulfur coal for steam power generation with air emissions of appreciably reduced environmental impact. When a sorbent material such as limestone or dolomite is fed with the coal into a combustion chamber maintained near 815°C, the sorbent reacts with the sulfur to produce a sulfate which can be removed and disposed with the solid ash materials. The resulting ash materials are discarded in a form more biologically available than ash from other combustion facilities. This characteristic justifies increased attention to the biological significance of the solid residues discharged as suspended particles in the flue gas, the particulate solids rejected from the system in bulk form and the leachates which might result from ground disposal of these particulate solids.

This EPA program studied, among others, a small demonstration pressurized FBC facility that burned coal with dolomite under pressure to generate heat for the production of steam plus several waste streams: (1) coarse particulates, (2) fine particulates, (3) gas, (4) second cyclone filter catch, and (5) bed reject material. Simple grab samples were taken of the coal, dolomite and waste streams 4 and 5 during the normal operation of the FBC facility. The SASS was used

Table II. Evaluation Criteria for Level 1 Biological Tests Used To Develop a Relative Ranking Based on Toxicity of Waste Streams

Biological Test	Units	Relative Sample Toxicity: High	Moderate	Low	Not Detectable
Ames assay					
Solid Samples	Lowest concentration[a] (mg/plate)	<0.5	0.05–0.5	0.5–5	>5
Liquid Samples	Lowest concentration[a] (μl/plate)	<2	2–20	20–200	>200
RAM and CHO					
Solid Samples	EC_{50} (μg/ml)	<10	10–100	100–1000	>1000
Liquid Samples	EC_{50} (μl/ml)	<6	6–60	60–600	>600
Rodent test	LD_{50} (g or ml/kg)	<0.1	0.1–1	1–10	>10
Aquatic tests	EC_{50}/LC_{50} (percent sample)	<20	20–75	75–100	>100
Plant stress ethylene	Lowest mixing time[a] (min)	1	2	7	>7
Seed germination–Root Elongation	EC_{50} (percent sample)	<0.01	0.01–1	1–100	>100
Insect toxicity					
Solid Samples	LD_{50} (mg/ml)	<0.5	0.5–5	5–50	>50
Liquid Samples	LD_{50}(μl/ml)	<5	5–50	50–500	>500

[a]Lowest concentration (Ames) or time (stress ethylene) giving a positive response.

to collect samples from waste streams 1 to 3. The samples were contained and transported according to level 1 procedures to the laboratory where level 1 biological and chemical analyses were performed.

The selection of the bioassays for use on the collected samples followed the Level 1 test protocol as described in Figure 1. Leachates were made of the large quantity solid streams—second cyclone catch, bed reject, coal and dolomite—for use in the aquatic tests. Since these solid streams, on exposure to or disposal in the environment, had the potential of leaching into both freshwater and marine habitats, both marine and freshwater species were used in the aquatic tests.

Health effects results from the FBC waste streams were observed only from the particulate samples when low levels of mutagenicity were found (Table III). The fine particulates also elicited a cytotoxic response. The rodent test detected no toxicity in any of the samples. In the ecological tests, coal leachate was toxic in all the aquatic tests (Table III). The bed reject leachate caused a toxic response in the freshwater, but not in the marine tests. Some toxicity was observed with second cyclone leachate (*Daphnia*) and the dolomite leachate (marine algae and invertebrates). No toxicity was observed in the plant stress ethylene test.

Level 1 chemical analyses were conducted on most of the FBC samples (dolomite and coal feedstocks excluded). Results were expressed as the ions or chemical classes which exceeded the discharge multimedia environmental goal (DMEG) value. This value is defined as the maximum concentration of a chemical believed to be safe for short term exposure to selected organisms [22]. DMEG values are derived primarily from laboratory animal studies. Using the chemical data expressed as those chemical constituents which exceed the DMEG value, comparison of the biological and chemical results can be made (Table IV). In general, there is positive correlation between the trends in biological toxicity and the chemical analyses with the gas sample being an exception. The lack of positive response in the plant stress ethylene test, even though several

Table III. Level 1 Bioassay Results for FBC Samples Including Toxicity Data and Relative Toxicity Evaluations[a]

Bioassay	Test Endpoint	Sample								
		Coarse Particulates	Fine Particulates	Gas	Second Cyclone	Second Cyclone Leachate	Bed Reject	Bed Reject Leachate	Coal Leachate	Dolomite Leachate
Ames	+/−	+ (L)	+(L)							
RAM	EC_{50} (μg or μl/ml)	>1000 (ND)	596 (L)		>1000 (ND)	>600 (ND)	>1000 (ND)	>600 (ND)		
Rodent	LD_{50} (g or ml/kg)				>10 (ND)	210 (ND)	710 (ND)	710 (ND)		
Freshwater										
Algae	EC_{50} (%)					>100 (ND)		45 (M)	19 (H)	>100 (ND)
Daphnia	LC_{50} (%)					92 (L)		41 (M)	71 (M)	>100 (ND)
Fish	LC_{50} (%)					>100 (ND)		25 (M)	92 (L)	>100 (ND)
Marine										
Algae	EC_{50} (%)					>100 (ND)		>100 (ND)	38 (M)	24 (M)
Invertebrate	LC_{50} (%)					73 (M)		>100 (ND)	66 (M)	75 (M)
Fish	LC_{50} (%)					>100 (ND)		>100 (ND)	72 (M)	>100 (ND)
Ethylene Stress (Plant)	Exposure (min)			>7 (ND)						

[a]H = high toxicity, M = moderate toxicity, L = low toxicity, ND = not detected (follows ranking scheme in Table II).

toxic constituents were present in the gas, may result in the insensitivity of the plant test (DMEG values are based primarily on animal data) to the chemicals present. Other considerations are possible antagonistic interactions among the constituents reducing the net toxicity of the sample, or deterioration of the sample during the elapsed time between collection and testing with toxicants adsorbing, precipitating, reacting or permeating through the walls of the sample container. The toxicity observed in the bed reject appears anomalous to the other samples in that it showed significant toxicity to all freshwater species and none to marine. This effect likely resulted from the high salt concentration which created osmotic imbalances in the freshwater species but had no effect on marine species tolerant of relatively high salt content. The FBC bed reject, coal and dolomite leachates, and particulate samples would all have high priority for level 2 testing. The gas and second cyclone samples would have lower priority for additional study.

LEVELS 2 AND 3 BIOLOGICAL TESTING

The development and refinement of level 1 bioassay protocols are quite advanced and they are now routinely implemented in EPA environmental assessment programs. The levels 2 and 3 bioassay protocols are still under development, both in concept and specific details. The basic objective of level 2 is to verify and elaborate on the level 1 results and to locate the probable cause of toxicity. For high-priority hazardous streams this will likely involve sample fractionation and subsequent testing. The fractionation procedure will likely involve separation of the stream into organic and inorganic components which would be tested. Further fractionation would be done on the toxic fraction(s): organics into acid, base, and neutral components and inorganics into cations and anions. These subfractions would also be tested. The testing of the fractions may use health and ecological bioassays similar to those employed at level 1 with additions such as increased number of replicates or test organisms which will improve the accuracy and reliability of the tests or a change in test species. New bioassays may be added such as teratogenicity, short-term chronic and life-cycle testing protocols. The objective of this scheme is to identify which component(s) of the waste system is toxic and to obtain information on the nature and magnitude of that toxicity.

Table IV. Comparison of Level 1 Biological and Chemical Results for Fluidized Bed Combustion Samples

Sample	Biological Results	Chemical Results
Coarse & Fine Particulates	Low mutagenicity, low cytotoxicity	Al,As,Be,Ca,Cd,Fe,Hg,K, Sb,Se,Si, aliphatic hydrocarbons >DMEG
Gas	No detectable toxicity	As,Cl,H_2SO_4,NO_x,Se,SO_2, >DMEG
Second Cyclone	Low to moderate toxicity only in aquatic invertebrates	Numerous elements >DMEG
Bed Reject	Moderate freshwater toxicity	Fe,Ni >DMEG, high salt concentration
Coal	Toxic in aquatic tests	No data
Dolomite	Moderate marine toxicity	No data

Level 3 will focus on process variations which result in changes to the chemistry of the waste streams. Flow-through testing may be employed where the concentrations to which the test organisms are exposed vary as the waste stream varies. Long-term chronic and subacute (nonlethal) testing may also be used to determine more subtle effects on the test species. This work will support the control technology development and implementation efforts.

Development of the levels 2 and 3 biological test protocol has not reached the pilot study stage. Considerable work remains to bring these activities up to the current stage of development for level 1. Importantly, results of level 1 are being applied in a variety of situations, including the textile manufacturing industry shown by Rawlings [23].

REFERENCES

1. Dorsey, J. A., L. D. Johnson, R. M. Statnick and C. H. Lochmüller. "Environmental Assessment Sampling and Analysis: Phased Approach and Techniques for Level 1," U.S. EPA Report EPA-600/2-77-115 (PB 268 563), Research Triangle Park, NC (1977).
2. Dorsey, J. A., L. D. Johnson and R. G. Merrill. "A Phased Approach for Characterization of Multimedia Discharges From Processes," in *Monitoring Toxic Substances,* D. Schueltzle, Ed. (Washington, DC: American Chemical Society, 1979), pp. 29–48.
3. Duke, K. M., M. E. Davis and A. J. Dennis. "IERL-RTP Procedures Manual: Level 1 Environmental Assessment Biological Tests for Pilot Studies," U.S. EPA Report EPA-600/7-77-043 (PB 268 484), Research Triangle Park, NC (1977).
4. Lentzen, D. E., D. E. Wagoner, E. D. Estes and W. F. Gutknecht. "IERL-RTP Procedures Manual: Level 1 Environmental Assessment (Second Edition)," U.S. EPA Report EPA-600/7-78-201 (PB 293 795), Research Triangle Park, NC (1978).
5. Ames, B. N., J. McCann and E. Yamasaki. "Methods for Detecting Carcinogens and Mutagens with the *Salmonella*/Mammalian-Microsome Mutagenicity Test," *Mutat. Res.* 31:347–364 (1975).
6. Brusick, D. J. "Bacterial Mutagenesis and Its Role in the Identification of Potential Animal Carcinogens," in *Carcinogens: Identification and Mechanics of Action,* A. C. Griffin and C. R. Shaw, Eds. (New York: Raven Press, 1979), pp. 95–107.
7. Ames, B. N. "Identifying Environmental Chemicals Causing Mutations and Cancer," *Science* 204:587–593 (1979).
8. Waters, M. D., D. E. Gardner, C. Aranyi and D. L. Coffin. "Metal Toxicity for Rabbit Alveolar Macrophages *in Vitro,*" *Environ. Res.* 9:32–47 (1975).
9. Garrett, N. E. "Development and Implementation of Cellular Toxicity Assays for Particulate Matter Using the Alveolar

Macrophage and Chinese Hamster Ovary Cell," EPA Contract No. 68-02-2566, Northrop Services, Inc., Research Triangle Park, NC (1980).

10. Mahar, H. "Evaluation of Selected Methods for Chemical and Biological Testing of Industrial Particulate Emissions," U.S. EPA Report EPA-600/2-76-137 (PB 257 912), Research Triangle Park, NC (1976).
11. Wininger, M. T., F. A. Kulik and W. D. Ross. "*In Vitro* Clonal Cytotoxicity Assay Using Chinese Hamster Ovary Cells (CHO-K1) for Testing Environmental Chemicals," *In Vitro* 14(4):381 (1978).
12. Sontag, H., N. Page and U. Saffiotti. "Guidelines for Carcinogen Bioassay in Small Rodents," NCI Technical Report Series No. 1, DHEW Publ. No. (NIH) 76-801, NCI-CG-TR-1 DHEW National Institutes of Health, Washington, DC (1976).
13. Balazs, T. "Measurement of Acute Toxicity," in *Methods in Toxicology*, G. Paget, Ed. (Philadelphia, PA: F. A. Davis Co., 1970), pp. 49–81.
14. Lewis, R. J., Ed. *Registry of Toxic Effects of Chemical Substances, 1978 Edition*, DHEW (NIOSH) Publ. No. 79-100, National Institute for Occupational Safety and Health, Cincinnati, OH (1979).
15. Committee on Methods for Toxicity Tests with Aquatic Organisms. "Methods for Acute Toxicity Tests with Fish, Macroinvertebrates and Amphibians," U.S. EPA Report EPA-660/3-75-009 (PB 242105), Corvallis, OR (1975).
16. Borthwick, P. W. "Methods for Acute Static Toxicity Tests with Mysid Shrimp (*Mysidopsis bahia*)," in *Bioassay Procedures for Ocean Disposal Permit Program*, U.S. EPA Report EPA-600/9-78-010 (PB 278 631), Gulf Breeze, FL (1978).
17. Miller, W. E., J. C. Green and T. Shiroyama, "The *Selenastrum capricornutum* Printz Algal Assay Bottle Test: Experimental Design, Applications, and Data Interpretation Protocol," U.S. EPA Report EPA 600/9-78-018 (PB 286 950), Corvallis, OR (1978).
18. Waterland, R. L., Co. "Terrestrial Ecology Protocols for Environmental Assessment Programs: Workshop Proceedings," U.S. EPA Report EPA-600/2-79-122 (PB 300 309), Research Triangle Park, NC (1979).

19. Tingey, D. C., C. Standley and R. Field. "Stress Ethylene Evolution: A Measure of Ozone Effects on Plants," *Atmos. Environ.* 10:969–974 (1976).
20. Santelman, P. W. "Herbicide Bioassay," in *Research Methods in Weed Science* (Champaign, IL: Weed Science Society of America, 1972), pp. 91–101.
21. Rubinstein, R., E. Cuirle, H. Cole, C. Ercegovich, L. Weinstein and J. Smith. "Test Methods for Assessing the Effect of Chemicals on Plants," U.S. EPA Report EPA-560/5/75-008 (PB 248 198), Washington, DC (1975).
22. Kingsbury, G. L., and N. D. Smith. "Update of EPA/IERL-RTP Environmental Assessment Methodology," Research Triangle Institute Report RTI/1914/22-01S, Research Triangle Park, NC (1980).
23. Rawlings, G. D. "Toward Improved Control of Toxic Substances," Chapter 7, this volume.

7

Toward Improved Control of Toxic Substances

Gary D. Rawlings
Monsanto Research Corporation
Dayton, Ohio

The previous two chapters have focused on the conceptual basis for understanding toxicological effects of chemical mixtures [1] and on a protocol for biological tests that identify which emission stream(s) are the most toxic [2]. This chapter extends these types of concepts and data into the world of control engineering. This chapter describes the events that spurred formation of a large program, the research methodology used to pursue it and the relative effectiveness of specific control technologies to reduce toxicity.

Two U.S. Environmental Protection Agency (EPA) regulatory events—the issuance of guidelines on effluent emissions in June 1974 and a court decree on effluent standards two years later—led to a large research effort designed to evaluate the toxicity of textile plant wastewaters and to determine the most effective treatment technology. By way of background the Effluent Guidelines Division of EPA set forth guidelines for the reduction of potentially toxic effluents through application of "best practicable control technology currently available" and "best available technology economically achievable" (BATEA). Issued in June 1974, the two types of reductions were to have been met by existing textile manufacturing point sources by July 1, 1977, and July 1, 1983, respectively [3]. In October 1974 the textile manufacturing industry, represented by the American Textile Manufacturer's Institute (ATMI), Northern Textiles Association, and Carpet and Rug Institute, filed a petition with the U.S. Fourth Circuit Court of Appeals asking for a review of the proposed 1983 effluent guidelines before implementation. Grounds for the suit were that the BATEA had not been demonstrated for the textile manufacturing industry. Following this, ATMI and EPA filed a joint motion for delay of the petition, stating that additional information would be developed through a cooperative grant study by ATMI and the EPA Industrial Environmental Research Laboratory (IERL), Research Triangle Park, NC.

The objective of the ATMI/EPA study was to gather enough technical and economic data to determine BATEA for reducing criteria pollutants from textile wastewaters. Selected pollutants and associated measurements for the textile industry included five-day biochemical oxygen demand (BOD_5), chemical oxygen demand (COD), chromium, color, pH, phenols, sulfide, and total suspended solids (TSS). The court instructed ATMI and EPA on January 3, 1975, to complete the study and review the results as quickly as possible.

To evaluate the best available technology, two mobile pilot plants were constructed and operated by Engineering Science, Inc. The mobile pilot plants allowed for real-time, flow-through treatability studies. Each pilot plant contained five tertiary wastewater treatment unit operations. One mobile unit was scheduled for studies at 12 textile plants and the other for studies at 11 plants. An additional tertiary treatment technology was laboratory tested.

Treatment operations in each mobile unit included a reactor/clarifier (using combinations of alum, anionic and cationic polyelectrolytes, and ferric chloride), dissolved air flotation capability, three granular activated carbon columns, and two multimedia filters, and ozonator capability. Powered activated carbon treatability tests were performed in the laboratory instead of in the field with the pilot plant. Using these six unit operations, ATMI and EPA selected seven treatment modes for evaluation, as identified below:

- A—reactor/clarifier → multimedia filter
- B—multimedia filter → granular activated carbon columns
- C—multimedia filter → ozonator
- D—ozonator
- E—reactor/clarifier → multimedia filter granular activated carbon → ozonator
- F—coagulation → multimedia filter
- G—dissolved air flotation

Mode E was optional. Each of the seven treatment modes was to be set up and operated individually over a two- to three-day period. Based on those data, the "best" system was to be selected and set up for two weeks of operational evaluation. Data from that evaluation were then to be forwarded for economic evaluation.

Prior to pilot plant field testing in the ATMI/EPA study, a second regulatory event occurred, which formed the basis for additional study—one based on toxicity measurements. On June 7, 1976, the U.S. District Court of Washington, DC issued a "Consent Decree" (resulting from *Natural Resources Defense Council et al.* v. *Train*) requiring EPA to accelerate development of effluent standards for 21 types of industrial point sources, including textile manufacturing. Among other requirements, the court's mandate focused federal water pollution control efforts on potentially toxic and hazardous chemical compounds. The consent decree required that "65 classes" of chemical compounds be analyzed in wastewater samples. Recognizing the difficulty of analyzing for all chemical species present in each class of compounds, EPA developed a list of 129 specific compounds representative of the classes listed in the consent decree. These compounds are referred to as "priority pollutants" and are divided into the following fractions for sampling and analytical purposes: asbestos, cyanide, metals, nonvolatile organics, pesticides, phenol, polychlorinated biphenyls and volatile organics. The EPA also developed a manual of sampling and analytical procedures to be used as a laboratory guide for the analysis of these priority pollutants [4].

The consent decree obligated EPA to identify which priority pollutants were present in industrial wastewaters and to determine the capability of various wastewater treatment technologies to remove priority pollutants. Therefore, EPA, with the cooperation of ATMI, decided to conduct another study to measure priority pollutants that would be separate from, but parallel to, the EPA/ATMI grant study. A contract for this new study was issued to Monsanto Research Corporation (MRC). Since the consent decree focused on the issue of wastewater toxicity, ATMI agreed to have samples collected for biological testing; this helped to assure a complete and comprehensive wastewater characterization database. The biological testing protocol developed by EPA for evaluating the reduction in toxicity of water samples by control technologies was integrated into the program [5].

OVERVIEW OF PROGRAM OBJECTIVES

The overall EPA textile program consisted of two separate projects, each with different activities, running parallel in time and converging toward the same goal: determination of the BATEA for textile wastewaters (Figure 1).

The EPA/ATMI study focused on identifying the BATEA for criteria pollutants. The MRC/EPA study emphasized the collection and analysis of samples for priority pollutant analysis and bioassay testing for the textile plant wastewater toxicity study; the fundamental objective was to determine the reduction in toxicity and priority pollutant concentrations achieved by the tertiary treatment technologies being tested at the 23 plants in the ATMI/EPA grant study.

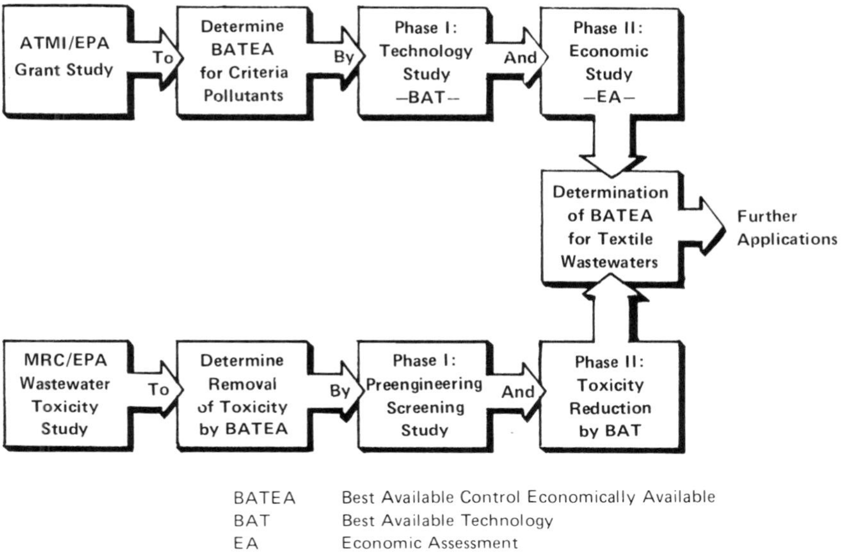

Figure 1. Overall program approach of two studies to determine BATEA.

The study, conducted at the MRC Dayton Laboratory, was divided into two phases to gather the most information in a cost-effective manner. Phase I was designed to collect baseline toxicity and priority pollutant data on the secondary treatment from the 23 plants before the pilot plant program began. In this manner, 23 samples were to be subjected to the battery of biological tests. Only those plants with toxic secondary effluents would be selected for measurement of reduction in toxicity by tertiary treatment systems. Also, appropriate biological tests could be selected instead of performing the entire battery of tests during phase II.

The purpose of phase II was to collect samples before and after each tertiary treatment unit operation and to compare any reduction in toxicity and priority pollutant concentrations at the plants selected from phase I. In addition to the collection of samples for priority pollutant analysis and biological testing, EPA included the new Level 1 environmental assessment methodology developed by the EPA Process Measurements Branch at Research Triangle Park [6]. Level 1, the first part of a three-phase environmental assessment approach, was designed to identify emissions that have a high potential for causing measurable health and ecological effects. In turn, available resources could be focused on the most toxic streams. As needed and based on the results of the level 1, approaches for levels 2 and 3 could be developed.

PHASE I: SCREENING STUDY

The purpose of the first phase was to collect chemical, physical and biological effects data from a variety of textile manufacturing sites. After the data were interpreted, further studies (phase II) followed.

Field Sampling and Analytical Methods

The basic textile plant wastewater treatment plant consists of an aeration lagoon with several surface aerators, followed by conventional clarifiers and chlorination. For phase I screening, raw waste and secondary effluent samples were collected at the points indicated in Figure 2. Secondary effluent samples were collected between the clarifier and chlorine contact basin, because that is the stream that would flow into a tertiary treatment system.

Raw waste samples were collected over an 8-hr period during a normal working day with automatic composite samplers. Eight individual secondary effluent samples were collected by grab sampling techniques using a 3-gal, Teflon®-lined, stainless steel bucket. Aliquots were removed from the bucket and poured into appropriate sample containers. Care was taken to ensure that the sample remained homogenous throughout each of the 10-min pouring sessions. Containers for volatile organics analysis were filled first and sealed to minimize possible evaporation losses. All samples were preserved in the field according to

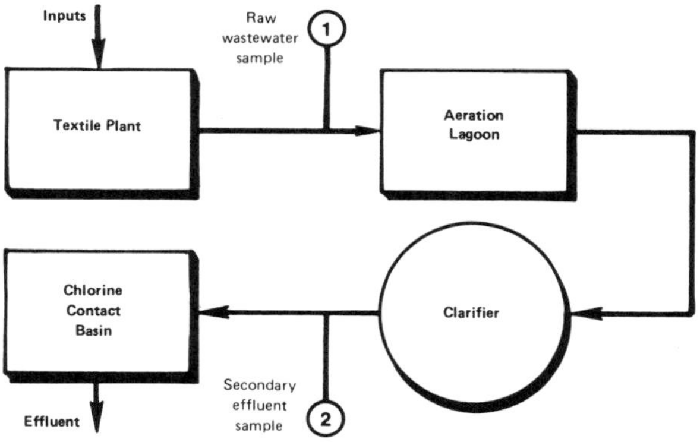

Figure 2. Overview of sampling locations for waste treatment study (phase II).

EPA specifications [7]. Samples were then packed in ice and shipped via commercial air freight to the appropriate laboratory for analysis.

Analysis of raw waste and secondary effluent samples (totaling 64 samples) for the 129 priority pollutants was performed by MRC in accordance with the analytical methodology recommended by EPA [4]. It is important to realize that the purpose of the EPA analytical scheme was to screen samples to determine which of the 129 chemical species were present. Also, the analytical results were reliable with concentration generally accurate to within a factor of two. (The analytical protocol was in the developmental stage, and further validation was necessary.) Regardless, the screening could result in a narrowed list of species; later, verification studies would more accurately quantify the concentrations of species.

The 114 organic priority pollutants were divided into four categories for analysis: acid extractable organics, base/neutral extractable organics, pesticides and Aroclors, and volatile organics [8]. Acid and base/neutral extractable organics were determined by extracting the sample with methylene chloride, first at pH higher than 11 and then the aqueous phase at pH lower than 2. Extracts were then dried on a sodium sulfate detected in a raw waste and secondary effluent sample were 14 and 8, respectively, with an average number per plant of 7 in the raw waste and 5 in the secondary effluent. The predominant compounds were *bis*(2-ethylhexyl) phthalate in 54 samples (with a range of 0.5–300 μg/l), toluene in 44 samples (0.4–300 μg/l), and ethylbenzene in 30 samples (0.7–3000 μg/l).

A summary of the 13 priority pollutant metals and cyanide concentrations in raw waste and secondary effluent samples is given in Table III. On an individual-plant basis it was frequently observed, especially for the metals data, that the concentration column and concentrated to 1 ml in a Kuderna-Danish evaporator with a Snyder column. Concentrates were analyzed using a Hewlett-Packard 5981 gas chromatograph/mass spectrometer (GC/MS) with a 5934 Data System. Pesticide and Aroclor species were similarly processed, but were extracted instead with a mixture containing 15% v/v methylene chloride and 85% hexane. Volatile organics were sparged from the sample with helium and adsorbed onto a Tenax column. Adsorbed species were later thermally desorbed for identification and quantification using the GC/MS system.

Since raw waste samples were collected with automatic samplers using a peristaltic pump and Tygon tubing, sample blanks were collected to determine if the sampler was contributing to the presence of organic priority pollutants. Laboratory-prepared, organic-free water was passed through the sampler and collected. Results of these analyses are presented in Table I.

Priority Pollutants Detected in Phase I

Results of GC/MS analyses of samples from 64 textile plant raw waste and secondary effluents for organic priority pollutants are provided (Table II). Of the 114 organic compounds in the priority pollutant list, a total of 44 different compounds as shown were identified in textile wastewaters. Raw waste samples contained 38, and 33 were in secondary effluent samples. On an individual-plant basis, the greatest number of organic compounds

Table I. Summary of Organic Priority Pollutants Found in Automatic Sampler Tubing Blanks

Fraction	Compound Found	Concentration Range (μg/l)
Acid	Phenol	0.6–1.1
Base/Neutral	Naphthalene	2
	Dimethyl phthalate	16
	Diethyl phthalate	0.5–10.2
	bis(2-Ethylhexyl) phthalate	1.5–46
	Di-*n*-butyl phthalate	1.3–1.7
Volatiles	Toluene	2.6–55
	trans-1,2-Dichloroethylene	3.2
	Trichloroethylene	2.4
	Ethylbenzene	8.3

of a specific pollutant was greater in the secondary effluent sample than in the raw waste sample. This phenomenon is due, in part, to the hydraulic retention time of the wastewater treatment facility. Since raw waste and secondary effluent samples were collected simultaneously, concentrations in the secondary effluent were due to raw waste loads that entered the treatment system 1–30 days prior to sampling. The average retention time for the 23 plants was about five days.

Chemical Analysis

Level 1 chemical analyses were performed on secondary effluent samples from 15 of the 23 basic textile plants. Level 1 protocol identifies classes of compounds present in environmental samples and measures the general concentration range. Results indicated that total concentration of methylene chloride extractable organics ranged from 3 to 64 mg/l.

In the level 1 procedure, each sample was fractionated by a liquid chromatography column into eight fractions based on polarity. Infrared analysis of each fraction indicated the presence of aliphatic hydrocarbons, aromatic compounds, esters, acids, fatty acid groups and phthalate esters. Low resolution mass spectrophotometric analysis of the eight fractions of each

Table II. Occurrence of Priority Organic Pollutants Combined from Raw Waste and Secondary Effluent Samples

	Number of Samples in Which Pollutant was Detected[a]			
Priority Pollutant	Total	Samples	Secondary Effluent Samples	Observed Concentration Range[b] (μg/1)
bis(2-Ethylhexyl) Phthalate	54	27	27	0.5–300
Toluene	44	22	22	0.4–300
Ethylbenzene	30	20	10	0.7–3000
Naphthalene	25	20	5	0.03–300
Diethyl Phthalate	21	12	9	0.2–70
Phenol	21	19	2	0.5–100
Chloroform	17	12	5	2–500
1,2,4-Trichlorobenzene	14	8	6	2–900
2,2-Dichlorobenzene	13	8	5	0.1–300
1,1,2,2,-Tetrachloroethylene	11	8	3	0.4–2100
Trichloroethylene	10	8	2	2–200
Acenaphthene	9	7	2	0.5–270
Di-*n*-butyl Phthalate	9	6	3	2–60
Pentachlorophenol	8	8	0	2–70
Dimethyl Phthalate	8	5	3	0.2–110
1,4-Dichlorobenzene	8	5	3	0.05–200
Chlorobenzene	8	6	2	1–300
Trichlorofluoromethane	8	2	6	2–2100
Benzene	6	4	2	0.5–200
1,1,1-Trichloroethane	5	5	0	2–300
Pyrene	5	1	4	0.1–0.9
Hexachlorobenzene	5	2	3	0.3–2
2,4,6-Trichlorophenol	3	2	1	0.7–20
N-Nitroso-di-*n*-propylamine	3	1	2	2–20
N-Nitrosodiphenylamine	3	1	2	2–20
Heptachlor	2	1	1	2–6
Antrhacene	2	1	1	0.1–4
Fluorene	2	2	0	5–15
1,1-Dichloroethane	2	2	0	0.6–4
cis-1,3-Dichloropropene	2	1	1	2–6
trans-1,3-Dichloropropene	2	0	2	0.9–4
2,4-Dimethylphenol	2	0	2	8–9
2-Chlorophenol	2	1	1	10–130
α-BHC	1	0	1	0.3
β-BHC	1	1	0	0.4
2,6-Dinitrotoluene	1	1	0	50
Indeno(1,2,3-c,d)pyrene	1	1	0	2
Butylbenzyl Phthalate	1	0	1	70
trans-1,2-Dichloroethylene	1	1	0	2
1,2-Dichloropropane	1	1	0	2
2-Nitrophenol	1	1	0	70
4-Nitrophenol	1	1	0	70
Chlorocresol	1	0	1	30
Bromodichloromethane	1	0	1	2

[a]Out of a total of 64 samples.
[b]Rounded to one significant figure.

Table III. Summary of Priority Pollutant Metals and Cyanide

	Concentration Range (mg/1)	
Element	Raw Waste Sample	Secondary Effluent Sample
Antimony	0.0005–0.06	0.0005–0.07
Arsenic	0.005–0.2	0.005–0.02
Beryllium	<0.0001	<0.0001
Cadmium	0.0005–0.05	0.0005–0.01
Chromium	0.0002–0.9	0.0002–2.0
Copper	0.0002–2.4	0.0002–0.3
Cyanide	0.004–0.2	0.004–0.2
Lead	0.001–0.2	0.001–0.2
Mercury	0.0005–0.004	0.0005–0.0009
Nickel	0.01–0.2	0.01–0.2
Selenium	<0.005	<0.005
Silver	0.005–0.1	0.005–0.1
Thallium	<0.005	<0.005
Zinc	0.03–8.0	0.07–38

sample detected the following types of compounds: alkyl phenols, azo compounds, *bis*(hydroxy-*t*-butyl phenol) propane, dichloroaniline, paraffinic/olefinic, toluene-sulfonyl groups, tri-*t*-butyl benzene and vinyl stearate.

Phase I Results of Biological Tests

The primary objective of the wastewater toxicity study was to determine the level of toxicity removal from secondary wastewater as achieved by the tertiary treatment technologies selected in the ATMI/EPA study. The purpose of this screening study was to provide chemical and toxicological baseline data on secondary effluents from the 23 textile plants and to select plants for the toxicity reduction study.

Biological tests were selected by EPA and included tests for assessment of both health and ecological effects [2,5]. Health effects tests estimated the potential mutagenicity, potential or presumed carcinogenicity and potential toxicity of the secondary effluent wastewater samples to microorganisms and mammals. Ecological effects tests focused on the potential toxicity of emission samples to vertebrates (fish), invertebrates (daphnids and shrimp) and plants (algae and beans)—representatives of freshwater, marine and terrestrial ecosystems.

Although toxicity tests with aquatic organisms can be conducted by applying wastewater samples directly to the test organisms, by injection or by feeding, most tests are conducted by exposing the test organisms to test solutions containing various concentrations of effluent samples. One or more controls are used to provide a measure of test acceptability by giving some indication of test organism health and the suitability of dilution water, test condition and handling procedures. A control test is an exposure of the organisms to dilution water with no effluent sample added. Biological tests are exposures of test organisms to dilution water with effluent samples added.

Biological testing, as well as chemical and physical parameters, should be considered when assessing the potential impact of industrial or municipal/industrial wastewaters on the aquatic environment. Biological testing involves determination of toxicity for samples of treated effluents. In an aquatic toxicity test, for example, aquatic organisms will integrate the synergistic and antagonistic effects of all the effluent components over the duration of exposure as explained by Anderson [1].

Eight biological testing systems were used for wastewater toxicity evaluation, utilizing many different organisms. Specific tests and the number of different indicator organisms used were: microbial mutagenicity with four indicator species and a total of 10 strains, cytotoxicity with two indicator cell lines, freshwater static bioassay with two indicator species, freshwater algal assay with one species, marine static bioassay with two indicator species, rangefinding acute toxicity using one species, and soil microcosm using a variety of soil organisms. Under guidance of appropriate EPA technical advisors, four of the eight biological tests were performed at five commercial laboratories, including MRC, experienced with the bioassays. The remaining four bioassays were performed by the EPA technical advisors.

A summary of the bioassay results is presented in Table IV, where toxicity is expressed as the percent of a secondary effluent solution that will cause the effect specified for each biological test over the testing period. For the cytotoxicity, daphnid and algal bioassays, an effective concentration 20 or 50 (EC_{20} or EC_{50}) was calculated. The EC_{20} for the cytotoxicity test means the concentration of secondary effluent that impairs metabolic processes in 20% of the test cells.

The viability test was a measure of the ability of the cells to survive exposure to the effluent and the adenosine triphosphate (ATP) test measured the quantity of the coenzyme ATP produced, indirectly measuring cellular metabolic activity. The EC_{50} for the algal tests means the concentration of secondary effluent that caused a 50% reduction in algal growth as compared to a control sample. The freshwater algae test was performed over a 14-day period and the marine algae test over a 96-hour period (see Duke and Merrill [2] for more details).

For the fathead minnow, sheepshead minnow and grass shrimp bioassays, death was used to measure toxicity, which was expressed as lethal concentration 50 (LC_{50}). Since rats were given a specific quantity of secondary effluent, toxicity was expressed as lethal dose 50 (LD_{50}).

The measure of toxicity to a soil microcosm was the quantity of carbon dioxide (CO_2) produced after effluent exposure as compared to a control sample. The quantity of CO_2 produced over a three-week period, after subtracting the quantity produced by the control, was plotted on graph paper. The slope of the curve then represented the rate of increase or decrease in CO_2 production due to exposure to the effluent.

Results of the freshwater ecology series showed enough variation in responses to permit relative ranking of the toxicity of effluent samples. However, no general rule can be formulated concerning the relative response between fathead minnows and *Daphnia*. For example, effluents of Plant E were found to be toxic to *Daphnia* but not to fathead minnows; the reverse was true at Plant T. For the marine ecology series, the data indicate that grass shrimp were more sensitive than sheepshead min-

Table IV. Summary of Biological Test Data for Secondary Effluent Wastewater Samples[a]

	Cytotoxicity		Freshwater Ecology Series			Marine Ecology Series			
Plant	Viability 24-hr EC_{20} (% Secondary Effluent)	ATP 24-hr EC_{20} (% Secondary Effluent)	Fathead Minnow 96-hr LC_{50} (% Secondary Effluent)	*Daphnia* 48-hr EC_{50} (% Secondary Effluent)	Algae[b] 14-day EC_{20} (% Secondary Effluent)	Sheepshead Minnow 96-hr LC_{50} (% Secondary Effluent)	Grass Shrimp 96-hr LC_{50} (% Secondary Effluent)	Algae[b] 96-hr EC_{50} (% Secondary Effluent)	Soil Microcosm Normalized Relative CO_2 Rate Change[c]
A	NAT[d]	NAT	19.0	9.0	76	62.0	21.2	e	−0.032
B	NAT	NAT	NAT	NAT	30	NAT	NAT	f	0.020
C	16.8[g]	6.1[g]	46.5	41.0	0[h]	69.5	12.8	90	−0.005
D	NAT	NAT	NAT	NAT	0[h]	e	e	e	−0.099
E	NAT	NAT	NAT	7.8	2[i]	NAT	NAT	10–50	−0.048
F	NAT	9.4	NAT	81.7	0[h]	NAT	NAT	85	−0.039
G	NAT	NAT	64.7	62.4	0[h]	NAT	NAT	59	0.018
H	NAT	NAT	j	k	96	e	e	e	−0.083
J	NAT	NAT	NAT	NAT	0[h]	e	e	e	−0.163
K	NAT	NAT	NAT	NAT	0[h]	NAT	NAT	77	−0.004
L	NAT	4.0	23.5	28.0	42	NAT	NAT	1.7	−0.020
M	NAT	NAT	NAT	60.0	0[h]	e	e	e	−0.059
N	13.3	3.8	48.8	1	2[j]	47.5	26.3	2.3	0.059
P	NAT	NAT	NAT	NAT	43	e	e	9.0	0.022
R[m]	NAT	NAT	16.5	8.0	93	e	e	e	−0.062
S	e	e	NAT	NSA[n]	0[h]	NAT	NAT	f	−0.017
T	NAT	2.5	46.5	NAT	0[h]	68.0	34.5	70	0.020
U	NAT	NAT	NAT	12.1	0[h]	NAT	NAT	f	0.055
V	NAT	NAT	36.0	9.4	0[h]	e	e	94	−0.066
W	NAT	13.7	55.2	6.3	94	37.5	19.6	50	0.031
X	NAT	4.8	NAT	NAT	0[h]	NAT	NAT	f	0.047
Y	NAT	NAT	NAT	NAT	0	e	e	e	−0.172
Z	NAT	NAT	NAT	42.6	18	e	e	e	−0.112

[a] No chemical mutagen was detected by the 10 microbial strains. No rat mortality after 14 days due to maximum dosage 10^{-5} mg/kg body weight (LD_{50}) however, five samples (B, F, L, N and S) showed potential body weight effects, and sample R resulted in eye irritation.
[b] Effect was algal growth inhibition.
[c] Negative sign indicated inhibition in CO_2 generation rate compared to a control sample; positive number indicates CO_2 stimulation.
[d] No observable acute toxicity.
[e] Analysis not performed on this sample.
[f] Growth inhibition less than 50% in 100% solution of secondary effluent.
[g] pH 9.1 not adjusted before testing.
[h] Sample stimulated algal growth.
[i] 95% growth inhibition in 2% solution of secondary effluent.
[j] Diseased batch of fish nullified this analysis.
[k] 40% dead at 100% concentration.
[l] 100% dead at all dilutions.
[m] Sample inadvertently collected prior to settling pond.
[n] No statistical analysis because heavy solids concentration obscured the analysis: sample did not appear to be acutely toxic.

nows. Also, fathead minnows were more sensitive, in the majority of the samples, than sheepshead minnows.

In terms of mutagenicity, none of the 23 effluent samples produced a positive response to any of the bacterial tester strains. The EPA Bioassay Subcommittee expressed some concern that the detection limits for this type of bioassay test were not sensitive enough to detect the presence of chemical mutagens in the 10- to 100-μg/l concentration range in environmental samples. Also, no acute toxicity was observed in any of the rat biological tests when rats were given the maximum dosage of 10 ml/kg of rat body weight.

Data Interpretation and Application

An objective of the screening study (phase I) was to rank textile plants according to the toxicity of their secondary wastewater and to select plants for detailed toxicity evaluation in phase II. To accomplish this objective, members of the EPA Bioassay Subcommittee met to evaluate the bioassay data (Table IV). Based on all of the bioassay data, the subcommittee ranked the 23 textile plants in descending order of secondary effluent toxicity (Table V). From these results, the subcommittee recommended that the following nine textile plants, ranked in relative order of toxicity, be tested to determine the reduction in toxicity achieved by the tertiary treatment technologies being tested in the ATMI/EPA grant study: N, A, W, C, T, V, L, S and P. (Plant R was also recommended for study in phase II because its secondary effluent samples were inadvertently collected prior to the settling pond.) In addition, the subcommittee recommended that the freshwater ecology series and Ames test be used to measure reduction in wastewater toxicity following treatment. The marine ecology series was not selected because none of the textile plants discharge wastewater into a marine environment.

There were several other significant results from the phase I toxicity screening. Of the 114 organic priority pollutants, only

Table V. Ranking of Textile Plants by Toxicity of Secondary Effluent

Toxicity Ranking	Plant
Most Toxic	N,A
↓	W
	C,T
	V,L
Least Toxic	S,P
Nontoxic	B,D,E,F,G,H,J,K, M,U,X,Y,Z

45 were detected in all samples. On an individual-plant basis, the largest number of organic species detected in a single effluent sample was eight, with an average number of five. Thallium was not detected in any of the samples. Thus, analysis of the data from the biological tests indicated it was possible to rank effluents based on relative toxicity.

PHASE II: VERIFICATION STUDY

The goal of the second phase was to measure the reduction in concentrations of specific toxic pollutants and the consequent reduction in toxicity resulting from tertiary treatment in the tertiary treatment systems of the mobile pilot plant. Figure 3 shows an overview of this trailer system. Only those plants selected in the first phase of the study were investigated.

Phase II Procedures

Sampling and analysis activities were conducted by MRC at 8 of the 23 pilot locations. At each location, the tertiary treatment technologies contained within the mobile pilot plants were arranged into different treatment systems. The systems selected were best suited to treat the secondary effluent generated by the existing biological (secondary) treatment plant present at each of the eight textile plants. In total, eight control technology systems were considered as follows:

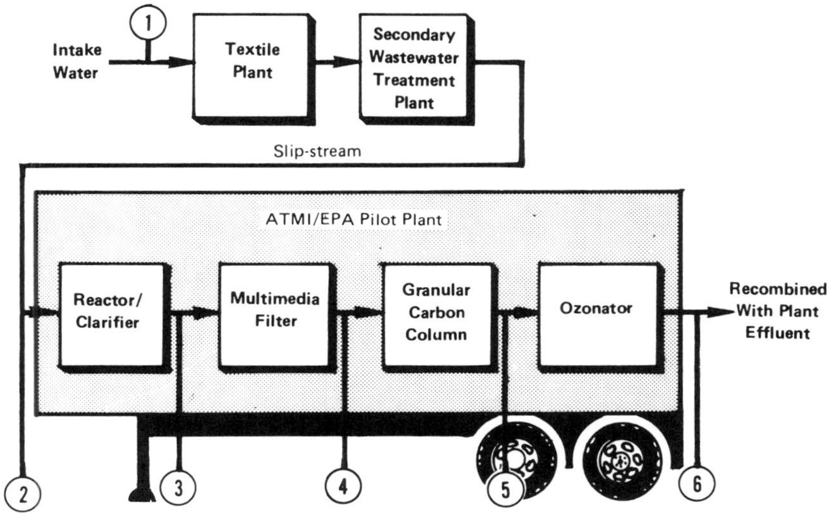

Figure 3. Overview of waste treatment processes (phase II) and sampling locations.

1. sedimentation;
2. coagulant addition → flocculation/sedimentation;
3. multimedia filtration;
4. coagulant addition → multimedia filtration;
5. coagulant addition → flocculation sedimentation → multimedia filtration;
6. multimedia filtration → granular activated carbon;
7. multimedia filtration → ozonation; and
8. coagulant addition → flocculation/sedimentation → multimedia filtration → granular activated carbon.

Table VI. Tertiary Treatment Systems Tested at Specific Pilot Plant Sites

	Type of Tertiary Treatment System Studied							
Plant	**1**	**2**	**3**	**4**	**5**	**6**	**7**	**8**
A		X			X			X
C		X			X			X
W	X		X			X	X	
S			X	X		X		
P		X	X		X	X		
N		X	X		X	X		
V			X	X		X	X	
T		X	X		X	X		

Table VI indicates which systems were operated at each plant [9].

Separate composite samples of textile plant intake water, secondary effluent and tertiary effluent were collected manually over a 24-hour period when the tertiary treatment systems under investigation at each plant were operated under steady-state conditions. Samples were chemically preserved, stored in ice at 4°C, and shipped by air freight or delivered by the sampling crew to the various laboratories for appropriate chemical analyses and bioassay studies.

The textile plant intake water, secondary effluent and tertiary effluent samples were analyzed by MRC for the 129 specific toxic pollutants. Analytical procedures followed those recommended by EPA [4]. The recommended analytical protocol, especially for toxic organic pollutant analyses, was still in the development stage and required further verification and validation. Consequently, the analytical results of textile wastewater samples were looked on as good estimates of which toxic pollutants were present, with concentrations accurate to within ±50%.

Bioassay studies were conducted on secondary and tertiary effluent samples. Two test species were used to test for mutagenicity (*Salmonella typhimurium* and *Escherichia coli*) [10–12], while a single cell line (CHO-K1) was used to test for cytotoxicity [13]. An EC_{50} value was obtained when freshwater algae were the test species. An LC_{50} was obtained when

Daphnia, fathead minnows and bluegills were the test species. Methods used in obtaining LC_{50} and EC_{50} values are those recommended by EPA for performing bioassay testing on pilot plant effluents [5].

Chemical analysis and biological test data were then assembled to select the treatment system that provided the best removal of toxic pollutants and toxicity. The data were also examined to identify other trends, as described in the next section.

Results of Chemical and Biological Tests

Seven toxic organic pollutants were found in the secondary effluent from at least one textile plant in concentrations greater than 10 μg/l (or the detection limit). These include *bis*(2-ethylhexyl) phthalate, 1,2-dichlorobenzene, 1,2,4-trichlorobenzene, toluene, methylene chloride, di-*n*-butyl phthalate and total phenols. However, *bis*(2-ethylhexyl) phthalate, methylene chloride and di-*n*-butyl phthalate found in the samples could have originated from contamination by the materials of which the pilot plant was constructed, or by sample analysis techniques. Except for mercury, selenium and thallium, all toxic inorganic pollutants were found in at least one textile plant secondary effluent in concentrations greater than 10 μg/l.

Of the eight secondary effluents submitted for algal bioassay, six were toxic (effluent demonstrated growth inhibition of greater than 50% when compared to a control), while five produced acute *Daphnia* toxicity in bioassay studies. Acute toxicity to fathead minnows was detected with only three secondary effluents of the seven tested. No mutagenicity or cytotoxicity was found in any of the secondary effluent or tertiary effluent streams tested.

Results of Control Strategy Tests

The analytical and bioassay data were used to identify positive and negative results that might be expected in applying various technologies to further treat secondary effluents from textile mills. Although additional research would be required to confirm the trends, there were, nevertheless, a series of statements that can be made about the demonstrated capabilities of the eight treatment systems to remove toxicity. They are summarized by types of problem and substances in the following paragraphs.

Acute Toxicity

Multimedia filtration and multimedia filtration followed by activated carbon adsorption appeared to be best for removing acute toxicity (as measured by bioassay tests). Intermediate removal ability was achieved by sedimentation, flocculation/sedimentation followed by multimedia filtration, multimedia filtration followed by ozonation, and flocculation/sedimentation followed by multimedia filtration and activated carbon. Least removal ability occurred with flocculation/sedimentation and multimedia filtration with precoagulation.

Cyanide

Cyanide was effectively removed by multimedia filtration followed by both activated carbon adsorption and ozonation. Multimedia filtration alone was ineffective in removing cyanide. Intermediate control was achieved by sedimentation, flocculation/sedimentation, multimedia filtration with precoagulation, flocculation/sedimentation followed by multimedia filtration, and flocculation/sedimentation followed by multimedia filtration and activated carbon.

Toxic Metals

Toxic metals were found to some extent in virtually all wastewater streams analyzed. Multimedia filtration, multimedia filtration followed by activated carbon adsorption and flocculation/sedimentation followed by multimedia filtration were most effective in reducing levels of toxic metals. Multimedia filtration with precoagulation and ozonation preceded by multimedia filtration were least effective. Intermediate removal utility was seen with sedimentation, flocculation/sedimentation, floccculation/sedimentation followed by multimedia filtration and activated carbon.

Toxic Organic Compounds

There were insufficient data on which to base conclusions. Toxic organic analyses yielded little information regarding relative effectiveness of the various tertiary treatment systems in reducing toxic organic pollutant concentrations since the secondary effluent from only two plants contained toxic organic pollutants, other than those that may have resulted from contamination, in concentrations greater than 10 μg/l.

Composite (Overall)

Best removal ability for the composite was multimedia filtration followed by activated carbon. Intermediate removal ability was seen with sedimentation, flocculation/sedimentation, multimedia filtration, flocculation/sedimentation followed by multimedia filtration, multimedia filtration followed by ozonation, and flocculation/sedimentation followed by multimedia filtration and activated carbon. Least removal ability was associated with multimedia filtration with precoagulation.

Several other trends in the data were identified. There appears to be a correlation between high residual coagulant concentrations in tertiary effluents and acute toxicity. High residual coagulant concentrations (greater than 9 mg/l) that result when coagulants used in a tertiary treatment system are improperly removed, had a detrimental effect, particularly upon algae and *Daphnia.* For example, Plant V was tested under one secondary and two tertiary treatment conditions. Residual iron was 0.21 mg/l in both the secondary and type 3 (multimedia filtration) tertiary treatment; this resulted in an algal LC_{50} (14-day) of greater than 100 for both conditions. By contrast, with type 4 (coagulant addition/multimedia filtration) tertiary treatment, the iron level was 6.2 mg/l and the algal LC_{50} (14-day) was 24.

It also appears that the presence of residual polymers in the effluent may be detrimental to algae. For example, the algal LC_{50} (14-day) for secondary treatment at Plant P was 53; by contrast, the algal LC_{50} for tertiary treatments that included addition of a polymer were 33 (multimedia filtration), 26 (coagulant addition/multimedia filtration) and 41 (coagulant addition and flocculation/sedimentation and multimedia filtration).

Third, algae appeared to be the most sensitive species used in biological testing. An EC_{50} (14-day) of less than 100% was observed in 74% of the samples tested. Acute toxicity to *Daphnia* was next at 54%, with other biological tests being below 44%. Therefore, algal bioassays should be considered first as a means to characterize the toxicity of textile mill wastewaters.

Summary of Phase II Research

The chemical analysis and biological test data obtained in the study were used to identify preliminary positive and negative results that would be expected in various control technologies to further treat secondary effluents from textile mills. Although the preliminary results obtained in this study will need to be confirmed, they are results that further guide plans to better manage toxic substances in our ecosystems. The following information was useful in that regard:

1. Only seven organic toxic pollutants at concentrations in excess of 10μg/l were seen in any of the secondary effluents of the eight textile plants. Three of the compounds may have resulted from contamination. But, overall, organic pollutant analyses did not provide a reliable basis for tertiary treatment system performance evaluation.
2. Eleven inorganic toxic pollutants were observed in at least

one of the eight secondary effluents in levels greater than 10 $\mu g/l$ (or the detection limit).

3. Of the eight tertiary treatment systems tested, multimedia filtration followed by granular activated carbon adsorption demonstrated the best overall toxic pollutant and toxicity removal capacity, as determined by inorganic toxic pollutant analyses and bioassays. Multimedia filtration with precoagulation demonstrated the least overall toxic pollutant and toxicity removal capability.
4. Tertiary treatment systems that left high levels of residual inorganic coagulant in their effluents generally increased the toxicity of the wastewater, as compared with treatment systems whose effluents contained lower levels. Also, systems employing certain cationic polymers used for coagulation increased the toxicity of the wastewater being treated, as measured by acute algal bioassays.

SUMMARY

After the need to manage potentially toxic effluents as identified by legislation, large programs were launched. Approaching the problem implicitly from a synergism/antagonism approach to chemical mixtures as explained by Anderson [1] and especially using the biological test protocol described by Duke and Merrill [2], the performance of various control strategies can be evaluated in a rather definite way. This was done during the ATMI/EPA and MRC/EPA studies of wastewater from textile manufacturing. Thus, information for selecting the BATEA was assembled and used to manage toxic substances in our ecosystems.

REFERENCES

1. Anderson, P. D. "Paradigms in Multiple Toxicity," Chapter 5, this volume.

2. Duke, K. M., and R. G. Merrill, Jr. "Development of New Biological Test Protocols," Chapter 6, this volume.
3. Gallup, J. D. "Development Document for Effluent Limitations Guidelines and New Source Performance Standards for the Textile Mills Point Source Category," U.S. EPA Report EPA 440/1-74-022a (PB 238 832), Washington, DC (1974).
4. "Draft Final Report: Sampling and Analysis Procedures for Screening of Industrial Effluents for Priority Pollutants," U.S. EPA, Cincinnati, OH (1977).
5. Duke, K. M., M. E. Davis and A. J. Dennis. "IERL-RTP Procedures Manual: Level 1 Environmental Assessment Biological Tests for Pilot Studies," U.S. EPA Report EPA-600/7-77-043 (PB 268-484), Research Triangle Park, NC (1977).
6. Hamersma, J. W., S. L. Reynolds and R. F. Maddalone. "IERL-RTP Procedures Manual: Level 1 Environmental Assessment," U.S. EPA Report EPA-600/2-76-160a (PB 257 850), Research Triangle Park, NC (1976).
7. "Manual of Methods for Chemical Analysis of Water and Wastes," U.S. EPA Report EPA-625/6-76-003a (PB 259 973), Cincinnati, OH (1976).
8. Rawlings, G. D. "Source Assessment: Textile Plant Wastewater Toxics Study, Phase I," U.S. EPA Report EPA-600/2-78-004h, Research Triangle Park, NC (1978).
9. Klieve, J. R., and G. D. Rawlings. "Source Assessment: Textile Plant Wastewater Toxics Study, Phase II," U.S. EPA Report EPA-600/2-79-019i, Research Triangle Park, NC (1979).
10. McCann, J., E. Choi, E. Yamasaki and B. N. Ames. "Detection of Carcinogens as Mutagens in the *Salmonella*/Microsome Test: Assay of 300 Chemicals," *Proc. Nat. Acad. Sci.* 72:5135–5139 (1975).
11. Ames, B. N., J. McCann and E. Yamasaki. "Methods for Detecting Carcinogens and Mutagens with the *Salmonella*/Mammalian-Microsome Mutagenicity Test," *Mutat. Res.* 31:347–364 (1975).
12. Slater, E. E., M. D. Anderson and H. S. Rosenkranz. "Rapid Detection of Mutagens and Carcinogens," *Cancer Res.* 31:970–973 (1971).
13. Wininger, M. T., F. A. Kulik and W. D. Ross. "*In Vitro* Clonal Cytotoxicity Assay Using Chinese Hamster Ovary Cells (CHO-K1) for Testing Environmental Chemicals," *In Vitro* 14(4):381 (1978).

8

Closing Remarks

Barney W. Cornaby
Health and Environmenal Sciences Section
Battelle Columbus Laboratories
Columbus, Ohio

Have we tamed the Medusa? We have made progress. Information presented by the participants did show that systematic identification and understanding of causes and effects are well underway. We learned that DDT adversely affects reproductive capacity in birds (Woodwell), chloroform causes cancer in rats (Reiches), cadmium and copper combined cause greater death rates in fish than either substance alone (Anderson), and complex mixtures from energy technologies and textile manufacturing plants harm algae and other aquatic life (Duke and Merrill; Rawlings). Also, we learned that various types of interventions to reduce the release and/or exposure to hazardous substances have been successfully implemented. In many parts of the world DDT has been banned or restricted in its uses, chloroform is undergoing examination for its reduced use, chromium and other heavy metals are being controlled to greater extent by newer changes in process streams, and complex chemical effluents also are being managed. Yet, as pointed out in the opening remarks there are thousands of substances whose potential toxicity is still unknown. Perhaps, a more correct answer to the above question is "Yes, we severed the heads of a few snakes, but there are many more with which to contend."

Many techniques and tools are being used to define problems and manage toxic substances. Some of these techniques and tools were mentioned by the participants: legislation (Woodwell), stepwise regressions (Reiches), microcosms and cesium-134 (O'Neill and Waide), formulas for concentration addition and response addition (Anderson), biological tests such as the Ames mutagenicity and algal LC_{50} assays (Duke and Merrill), and multimedia filtration and ozonation (Rawlings). These tools should be used in the correct order, just as a surgeon must use correct tools in a prescribed order. Dr. Samuel J. Kiehl, III (personal communication), explained that in a carefully performed amputation, scalpels would be used to open the skin, scissors such as Deaver or Kaylor would be used to cut coarse tissue, with the Iris, Mayo, Metzenbaum or similar scissors used to dissect delicate tissues, followed by an osteotome to saw the bone. Similarly, a hatchet job on the toxic substance problem will not suffice; instead a more methodical, tedious approach to its management will be taken.

On the first day of the symposium, two powerful perspectives—the human and the ecosystem—were advanced. That they are different is accentuated by noting selected terms used in the respective disciplines of each. From the human perspective, words such as pharmacokinetics, hepatomas, lung cancer rates, Broad Street pump and carcinogenic risk bring to mind the vast resources committed to health care, hospitals and the prolongation of human life independent of the natural ecosystems. From the ecosystem perspective terms such as cycling in a mixed deciduous forest, hierarchy theory, pH-dissolved O_2 phase-space in aquatic microcosms, defoliation and trophic dynamics conjure images of analyzed vistas of forest, air masses and lakes exchanging nutrients and energy—vistas that sometimes do not include humans. The human perspective often emphasizes a view of man and nature. Focus is on the internal milieu that is manipulated by the introduction of drugs and other chemicals. The ecosystem perspective often advances a man in nature view. Most manipulations in this external milieu result in changes to soil, plants and animal relationships. Furthermore, the human perspective concentrates on one important part of the world, while the ecosystem perspective embraces all parts. Enough of the reminders.

Each perspective contributes its own approaches, techniques and tools to solve toxic substance problems. Yet, many toxic substance problems are common to both perspectives. Human-oriented effects research that uses the principles of ecology and ecological effects research that pays more attention would surely better use the best of each. Two practical ways for such a synthesis come to mind: (1) use nonhuman organisms in ecosystems as early warning measures to protect humans and (2) intervene in biogeochemical cycles to reduce the flow of carcinogens to humans. Implementation of both would dull the cutting edge of dangerous environmental agents to man. Each will be treated in the following paragraphs.

First, one practical mechanism for combining both perspectives is to expand the use of nonhuman organisms to protect humans. This is an old practice whose time has come again. Canaries and dancing mice are more sensitive to carbon monoxide and other noxious gases than are humans. Years ago, miners noting loss of activity or death of one of these feathered or furred indicators were alerted that poisonous gases were present in the drift or shaft. The miners would then be warned to ascend rapidly to ground level and safety. In the case of Minamata disease, human illness was presaged by octopi and fish floating dead near shore, birds crashing into the sea and rocks, and poisoned cats undergoing paroxysmal fits—all early warnings of the danger of methylmercury pollution, a danger that eventually reached and harmed humans living in the nearby villages. Such dyads, an effect on a sensitive organism followed by an effect on a less sensitive organism, have not systematically been recognized and cataloged. True, we have employed the sense of this approach for decades in laboratory testing and clinical studies, but such relationships have not been worked out for real-world or field settings. Once an ecosystem or ecosystem component effect(s) were defined as well as are those for humans, we could launch the necessary companion studies. Ecosystem and epidemiological studies conducted simultaneously should identify workable dyads. Then, hopefully, sensitive ecosystem effects alone, such as nutrient release patterns in microcosms, could be used to warn impending changes in human morbidity and mortality.

A second way to solve toxic substance problems using parts of both the human and ecosystem perspectives is to document biogeochemical pathways and cycles of materials that are known or suspected human carcinogens. Exposures to certain forms of arsenic, chromium, nickel and selenium may cause tumorigenic responses in humans. While a few pathways from a source to man are simple, i.e., movement in air, the study of the more complex pathways from the total system and biogeochemical cycling viewpoints are not common in epidemiologic studies. The ecologist could contribute greatly here to round-out the knowledge needed to understand the conditions of various flows of harmful materials to humans, i.e., donor- or recipient-controlled transfers. In other words, the study of human carcinogens cycling in a forest, lake or urban ecosystem should reveal the storage compartments and transfer patterns of particularly dangerous materials. With this knowledge we could better intervene at specific locations and times in the ecosystems to reduce cancer.

Chemical mixtures and their effects received considerable attention in the symposium. All presentations mentioned them, and three (Anderson, Duke and Merrill, and Rawlings) dealt with them directly. Anderson sharpened our perception of multiple toxicity, using many examples to show us the unexpected results associated with chemical mixtures. We came to believe that the carefully performed multiple toxicity approach, as opposed to the single chemical species approach or to an all-encompassing toxic unit formula, was the best way to proceed with the setting of standards for management purposes. Anderson's and similar work pointed to the needs for implementation of bioassay systems that would supplement the strictly compound-specific chemical analyses. Duke and Merrill revealed the structure and purposes of such a biological testing protocol that uses both health and ecological effects tests. The protocol works. One can clearly identify from the toxicity viewpoint which waste streams need more control attention, as illustrated in their example using fluidized-bed effluents. Real closure came with Rawlings's paper. He extended the application of bioassays (that use mixtures rather than single chemical species) into the world of control technology. Control technology, guided by results from biological and chemical tests, provides the practical tool for managing the introduction of toxic substances into our ecosystems. A good example was provided with the waste from textile manufacturing.

Many potentially toxic substances are beneficial—when properly used they enhance agricultural production, extend life and provide more energy. On the other hand, regulation, control, amendments to laws and new laws have their toll. Hastily applied or indiscriminate environmental legislation not only will kill the incentive to create, produce and market new chemical products, but may confer little additional protection to ourselves and our ecosystems. These and related issues arose during the symposium discussion (see Preface for access to this dialog). Indeed, the toxic substance issues as George M. Woodwell pointed out, "embody nearly all the elements of the contemporary crisis of environment; growth, profits, and economic and political power." As we work to better manage toxic substances in our ecosystems we will face and reface inherent conflicts. To achieve balance demands creative and timely input from thousands of competent people over the coming decades.

Author Index

Addy, N. D. 57,72
Alabaster, J. S. 99
Allen, T. F. H. 47,53,60,68,73
Ames, B. N. 118,144
Anderson, J. M. 71
Anderson, M. D. 144
Anderson, P. D. 4,75,80,96,98, 132,143,145,146,148
Andrew, R. W. 98
Aranyi, C. 118
Ashby, W. R. 52,53,71
Ausmus, B. S. 69,71,72

Baas-Becking, L. G. M. 50,69
Balazs, T. 119
Banathy, B. H. 68
Bartell, S. M. 72,73
Beamer, P. 99
Beamish, R. J. 97
Behnke, J. A. 69
Beidleman, R. G. 72
Beijer, K. 99
Bement, R. E. 72
Benz, S. L. 67,69
Bingham, E. 94,99
Blackburn, T. R. 68
Bliss, C. I. 82,97
Blot, W. J. 40
Bormann, F. H. 69,70
Borthwick, P. W. 119
Bosserman, R. W. 68
Boucot, K. R. 41
Broderius, S. J. 99
Brown, C. C. 40
Brown, E. R. 99
Brusick, D. J. 118
Buncher, C. R. 41
Burbank, F. 40
Burk, R. F. 98
Butler, G. C. 96,99

Cantor, K. P. 32,41
Caswell, H. 55,71
Chan, Y. K. 98
Chen, A. 97
Chew, R. M. 71
Chi, P. 41
Choi, E. 144
Chudyk, W. 97
Chung, J. 40
Clarke, P. J. 67,69
Clarkson, S. P. 72

Clarkson, T. W. 97
Cleland, J. G. 120
Coffin, D. L. 118
Cole, H. 120
Conrad, M. 67
Cornaby, B. W. 1,71,145
Correll, D. L. 70,72
Coté, M. G. 99
Cottam, C. 17
Crossley, D. A., Jr. 70–72
Cuirle, E. 120
Czyrska, H. 98

d'Apollonia, S. 80,96,98
Davis, M. E. 118,144
DeAngelis, D. L. 53,71
de Luca, J. 98
Dennis, A. J. 118,144
Dick, J. 98
Dix, R. L. 72
Dorsey, J. A. 118
Douglass, J. E. 70,72
Duke, K. M. 4,94,99,101,118,
133,143–146,148,149
Durham, S. L. 45,68

Eaton, J. S. 70
Edwards, N. T. 71
Elcombe, C. R. 99
Elton, C. S. 71
Elwood, J. W. 70
Emanuel, W. R. 60,70,73
Ercegovich, C. 120
Eschenbrenner, A. B. 31,41
Estes, E. D. 118
Ettinger, M. B. 40
Evans, F. C. 67

Field, R. 120
Finn, J. T. 68
Finney, D. E. 97
Fisher, D. W. 70
Foster, K. A. 98
Franzin, W. G. 97
Fraumeni, J. P. 40
Freedman, M. L. 67,69

Gale, W. G. 68
Gallepp, G. W. 72
Gallimore, B. 96,97
Gallup, J. D. 144
Gardner, D. E. 118
Gardner, M. R. 52,53,71
Gardner, W. S. 40
Garrett, N. E. 118
Gentry, J. B. 71
Ghent, A. W. 72
Giddings, J. M. 59,72
Giesy, J. P. 67–69
Gist, C. S. 71
Grant, M. C. 97
Green, J. C. 119
Greenfield, P. M. 98
Greenspan, I. 99
Grier, C. C. 70
Griffin, A. C. 118
Gutknecht, W. F. 118

Hains, J. J. 67,69
Halfon, E. 68
Hamersma, J. W. 144
Hammond, H. C. 40
Harris, R. H. 32,41
Harris, W. F. 69
Hasler, A. D. 70
Hazdra, J. S. 99
Heivitt, A. 96
Henderson, G. S. 69,70
Herbert, D. W. M. 86,98
Hewitt, L. A. 97

Hewitt, W. R. 99
Hiatt, H. H. 40,41
Higgins, H. 17
Hill, J., IV 45,68
Hines, N. O. 16
Hodson, P. V. 98
Hoel, D. G. 41
Hogan, M. D. 32,41
Hoover, R. 40,41
Horm, J. M. 39
Horovitch, H. 96
Howell, F. G. 71
Humpstone, C. C. 5,16
Hurd, L. E. 59,72
Hutchinson, G. E. 49,68
Hutchinson, T. C. 97,98

Iltis, H. H. 53,68
Innis, G. S. 72

Jackson, D. R. 69
Jernelöv, A. 94,99
Johnson, L. D. 118
Johnson, N. M. 69,70
Johnson, P. L. 70
Junk, W. 69

Kaplan, I. R. 69
Kappas, A. 94,99
Keith, L. 99
Kellerman, G. 42
Kingsbury, G. L. 120
Kiper, J. P. 98
Kitchell, J. F. 72
Kloepfer, R. D. 40
Klieve, J. R. 144
Koenig, H. E. 71
Konasewich, D. E. 97,98
Koonce, J. R. 72,73
Krebs, J. E. 72
Kruse, C. W. 41
Kulik, F. A. 119,144
Kuschner, M. 40
Kuzma, C. M. 41
Kuzma, R. J. 41
Kwapinski, J. B. G. 99

Lane, P. A. 67
Laszlo, E. 68
Lauff, G. H. 67
Lech, J. J. 99
Lentzen, D. E. 118
Levin, S. A. 67
Levins, R. 67
Levy, A. 40
Lewis, R. J. 119
Lewis, W. M., Jr. 97
Lexon, P. L. 98
Likens, G. E. 69,70
Lindeman, R. L. 67
Lloyd, R. 86,98
Lochmüller, C. H. 118
Loucks, O. L. 69
Lowe-McConnell, R. H. 69
Luyten-Kellerman, M. 42

Macek, K. J. 99
MacFadyen, A. 71
Maddalone, R. F. 144
Mahar, H. 119
Maines, M. D. 94,99
Marking, L. L. 99
Mason, T. J. 40,41
Mattson, W. J. 57,72
Mauck, W. L. 99
May, R. M. 59,73
McCann, J. 118,144
McFarlane, G. A. 97
McKay, F. W. 40
McNaughton, S. J. 72

Mellinger, M. V. 72
Menzie, G. M. 17
Merrill, R. G., Jr. 4,94,99,101, 118,133,143–146,148,149
Miller, E. 31,41
Miller, W. E. 119
Monk, C. D. 70
Moore, D. 69
Morgan, J. J. 69
Morowitz, H. J. 69
Myslik, G. 97

Niemeier, R. W. 99

Odum, E. P. 67
Odum, H. T. 67
Olsen, K. 42
O'Neill, R. V. 4,43,52,59,67,69, 70,72,146
Overton, W. S. 67
Owen, D. G. 72

Page, N. 41,119
Page, T. 41
Paget, G. 119
Passow, H. 97
Pattee, H. H. 68,71
Patten, B. C. 48,54,57,59,67–69, 71
Peeler, J. T. 40
Perry, S. 98
Pettigrew, D. P. 67,69
Pierce, R. C. 98
Pierce, R. S. 69,70
Plaa, G. L. 99
Pollack, E. S. 39

Rall, D. P. 40
Rawlings, G. D. 4,118,120,121, 144–146,148,149
Reichel, W. L. 17
Reiches, N. A. 4,19,41,65,73, 145,146
Reichle, D. E. 59,67,69
Reid, J. B. 99
Reynolds, S. L. 144
Ridley, W. P. 97
Roberts, A. 53,71
Rook, J. J. 40
Rosen, A. A. 40
Rosenkranz, H. S. 144
Rosen, R. 45,68,72
Ross, Q. E. 71
Ross, W. D. 119,144
Rothman, K. J. 40
Rothstein, A. 97
Rubinstein, R. 120

Saalfeld, E. E. 40
Saffiotti, U. 41,119
Salt, G. W. 67
Santelman, P. W. 120
Saunders, R. A. 40
Schafer, M. L. 28,40
Scheiman, M. A. 40
Schindler, J. E. 46,51,67,69
Schissel, L. A. 67,69
Schneiderman, M. A. 40
Schnell, R. C. 98
Schreiner, S. P. 67,69
Schueltzle, D. 118
Seba, D. B. 97
Segall, H. J. 97
Selikoff, I. J. 40
Setzler, E. M. 72
Shaw, C. R. 42,118
Shiroyama, T. 119
Shugart, H. H., Jr. 70,73
Shurben, D. G. 86,98
Simon, H. A. 48,67,68

Sinnhuber, R. O. 99
Skeel, R. T. 40
Sladen, W. J. L. 17
Slater, E. E. 144
Smith, J. 120
Smith, L. L. 99
Smith, M. H. 71
Snell, F. M. 72
Soderlund, L. 99
Sontag, H. 119
Spear, P. A. 96,98
Sprague, J. B. 97
Standley, C. 120
Starr, T. B. 47,68
Statnick, R. M. 118
Stokes, P. M. 97
Stumm, W. 69
Swank, W. T. 70,72

Tansley, A. G. 66
Taub, F. B. 71
Tedeschi, D. H. 40
Tedeschi, R. E. 40
Tennis, P. S. 72
Thayer, J. S. 97
Tingey, D. C. 120
Todd, R. L. 70,72
Tregonning, K. 53,71

van Dobben, W. H. 69
Van Hook, R. I. 69
Van Loon, J. C. 97
Van Voris, P. 52,60,69,70
Veith, G. D. 97

Wagoner, D. E. 118
Waide, J. B. 4,43,45,51,53,59, 60,67–70,72,146
Waldichuk, M. 97
Waldron, M. C. 67,69
Wales, J. H. 99
Waring, R. W. 67,70
Warren, C. E. 96
Washburne, C. 69
Waterland, R. L. 119
Waters, M. D. 118
Watson, A. P. 69
Watson, J. D. 40,41
Webb, D. 72
Weber, L. J. 80,96
Webster, J. R. 45,53,68,70
Weinberg, G. 67
Weinstein, L. 96,120
Weinstein, N. 98
Weiss, P. A. 46,68
Weiss, W. 41
West, D. C. 73
Whitby, L. M. 97
Whittaker, R. H. 67
Wiegert, R. G. 68,72
Wiens, J. A. 67
Wilkins, J. R. 41
Wininger, M. T. 119,144
Winsten, J. A. 40,41
Witkamp, M. 57,69,71
Wolf, L. L. 72
Wong, P. T. S. 93,98
Wood, J. M. 97
Woodwell, G. M. 3,5,57,67,69, 72,145,146,149

Yamasaki, E. 118,144

Subject Index

absorption 75,101
abundances 60
acenaphthene 130
acid 131
  precipitation 81
acute exposures 76,108,111
acute toxicity 110,133,135,139–142
additive combinations 95
additive effect 79
additive mechanism 81
additive paradigms 76
additive response 88
additivity 81,93,95
adenosine triphosphate (ATP) test 109,133,134
aeration lagoon 126,127
aerial transport 12
age 83
agencies, regulatory 20,39
agricultural production 3,149
agriculture 11,14
air
  emissions 112
  masses 146
  movement in 148
  pollution 6
Alberta Oil Sands Environmental Research Program 96
alcohol(s) 21,29
aldrin 28
algae 51,88,93,110,114,115,132,134,138,141,142,145
  bioassays 133,139,142,143
  growth 110
  growth reduction 133
  LC_{50} assay 146
  production 58
  test protocol 110,133
  toxicity test 108
aliphatic hydrocarbons 29,117,131
aliquots 126
alkalinity 90
alkyl phenols 131
Alsophila pometaria 57
aluminum 117
ambient concentration 78,89,90
ambient exposure 33
ambient levels 81,90

American Chemical Society 6
American population, high mobility of 34
American Society of Testing Materials (ASTM) 106
American Textile Manufacturing Institute (ATMI) 122–124
/EPA grant study 122–127, 132,135,143
Ames assay 106–109,113,114, 136,146
amphibians 12
anabolic process 45,50
analytical methodology 127
analytical protocol 127,138
analytical purposes 124
aniline 23
animal(s) 106
assays 23
body 53
data 116
models 24
relationships 147
species 22
studies 20,21,25,114
tests 22
whole 109
anions 116
antagonism 79,87
antagonistic agent 87
antagonistic effect 4,24,132
antagonistic interactions 103, 116
antagonistic role 86
anthracene 130
anthropogenic alterations 52
antimony 117
Appalachians, southern 60
aquatic environments 81
aquatic life 145
aquatic microcosms 51,60,146
aquatic organisms 88,94,132
aquatic populations 81
aquatic tests 105–106,110,113–114
aquatic toxicity 107–108
aquatic weeds 88
aqueous chlorine species 29
aqueous leach procedure 106
Aroclors 128
aromatic compounds 131
arsenic 21,22,117,148
arthropods 11,56–57
artificial assemblage 54
aryl hydrocarbon hydroxylase (AHH) 38
asbestos 25,124
assays
See Ames assay; bioassay
assemblage, artificial 54
assimilative capacity 7,13–15
Athena 2
atmosphere 9,12,15
atmospheric patterns 9
atmospheric transport 10
Atomic Energy Commission 9
Australia 54
automatic sampler(s) 126–129
automobile exhaust 28
autotrophic organisms 51
availability 15
azo compounds 131

bacteria 108
soil nitrifying 57
bacterial colonies 109
bacterial tester strains 135
balance, achieving 149
ban (a substance) 39
Basel, Switzerland 11
baseline 51

data 132
toxicity 125
beans 132
bed reject leachate 112–117
behavior
dynamic 53–57
frequency 47–48
holistic 44
mode 48
unconstrained 54
benzene 22,130
benzopyrene 23
benzo[a]pyrene (B[a]P) 25
3,4-benzpyrene 28
best available technology (BAT) 125
best available technology economically achievable (BATEA) 122,124,143
α-BHC 130
β-BHC 130
Bikini Atoll 8
binary mixtures 77,83,87–90,93
binding sites, total available 77
binomial model 34
bioaccumulation 57,66
bioassay 114,115
acute 103
algal 133,139,142,143
Ames 106–109,113,114,136, 146
animal 23
chronic 103
daphnid 133
data 135
ecological 116
experimental 22
health 116
protocols 101,116
results 112
static 103,133
studies 138–139
systems 149
test 124,135,139,140
Bioassay Subcommittee 104
biochemical effects 77
biochemical oxygen demand (BOD_5) 122
biogeochemical cycles 4,50, 147–148
biogeochemical pathways 50, 148
biogeochemical processes 51
biogeochemical properties 63, 64
biogeochemical responses 52
biogeochemical state 51
biogeochemical systems 46,49, 51
biogeochemical variables 64
biological analysis 104,107,114
biological effects 88,126
biological hierarchy 53
biological mechanisms 24,36
biological organization 66
biological populations 57
biological results 117
biological systems 45
biological test(s) 4,105–107, 113,121–125,131,132,142,146
biological test data 139
summary 134
biological test protocol 117,133, 143,149
biological toxicity 114
biological treatment plant 137
biomass, total 60
biota 7,10,13,15
biotically active substances 15
biotic components 60,62
biotic effects 7,8,11
biotic evolution 7

biotic hazard 10
biotic interactions 54,55
biotic molecules 10
biotic systems 7,8,12–15
biotic variables 89
bird(s) 145,147
  carnivorous 12,57
  populations 13
birth defects 39
Blackstone (lawyer) 5,16
bluefish 14
bluegills 139
body size 22,23
bombs
  fission 8
  tests of 8,9
boreal forest 11
bound sites 79
"boundary conditions" 51
boundary membrane 88
Brachydanio rerio 83
BRAVO 8
Broad Street pump handle 36, 146
brominated compounds 29
bromine 29
bromodichloromethane 30,130
bromoform 30
budgets, net 51
bushbean 110
butylbenzyl phthalate 130

cadmium 21,52,81–95,117,145
  -mercury mixture 82–86,95
  -zinc mixture 82–86,95
calcium 117
  carbonate 91
  cycles 59
  ions 89
canaries 147
cancer 3,19,21,25,26,39,94,145, 148
  biology of 21
  bladder 32,35
  breast 20
  colon 20,32
  esophagus 32
  gastrointestinal tract 31
  human 4,21,22
  incidence 20,23,34
  large intestine 35
  lung 20,25,26,36,38,146
  mortality rates 6,20,26,27,31
  pancreas 32
  pathogenesis of 25
  rates 25,27
  respiratory tract 25
  -sensitive 63
  skin 63
  stomach 32
  urinary tract organs 31
  See also carcinogens; carcinogenesis; carcinogenic effects; carcinogenicity
cankerworm, fall 57
carbamates 93,94
carbon 50
carbon dioxide 133
  efflux 52,63,66
  production 135
carbon monoxide 147
carbon tetrachloride 30,94
carcinogen(s) 19,21,22,23,24, 25,27,28
  animal 21,31
  chemical 22,109
  control of 38
  environmental 19–42
exposure to 20

human 21–23,147,148
potential 23,109
See also cancer
carcinogenesis
chemical 94
environmental 96
processes 19
risk 146
See also cancer
carcinogenic effects 21
carcinogenicity 30,31,36
potential 132
presumed 106,132
carcinoma 20,31,32
carnivores 64
Carpet and Rug Institute 122
case-control study 24
catabolic process 45,50
catastrophe 5
cations 116
causality 20
causal links 64
cause-effect relationships 20
cell(s)
division rates 22
susceptible 23
viability 109
cellular level 108
cellulase activity, net 64
centers
agricultural 76
industrial 76
municipal 76
cesium-134 146
cesium-137 10
chemical(s) 6,19,23,37,39,63, 116,147
analyses 114,129,138,139,142
anthropogenic 44
bond energy 50
carcinogenic 22,109
change 15
classes 103
composition 14,15
compounds 21,123
compounds, analyses 149
compounds, registered 2
constituents 28,93
contaminants 76
data 4,112
dumps 6
effects 126
effluents, complex 145
environment 21
environmental agent 77,147
equilibrium analysis 50
hazardous 103,145
mixtures 4,88,107,121,143, 148,149
new 27
pollutants 75,94
potential, low 50
production 26
results 117
species, single 148,149
tests 149
toxic 37
waste disposal 3
chemical oxygen demand (COD) 122
chestnut, American 60,61
chestnut blight 54
Chinese hamster ovary (CHO) test 107,108,109,113
chlordane 28
chlorinated hydrocarbons 14, 43,57
chlorination 29,30,33,126
chlorine
aqueous species 29

-carbon ratio 29
contact basin 126,127
dose 33
chlorocresol 130
chloroform 30–33,94,130,145
bis(2-chloroisopropyl) ether 29
bis(chloromethyl) ether 36
2-chlorophenol 130
cholera 36
Christian doctrine 16
chromium 21,122,145,148
circular causal systems 49
circulation
atmospheric 9
oceanic 9
clarifier 127
conventional 126
classical systems approach 45
Clean Air Act 37,101
Clean Water Act 101
closed systems 16
coagulant addition 137,142
coagulation 123
coal 114,117
burning of 112
high-sulfur 112
leachate 114–116
cobalt 94
cocarcinogens 94
coenzyme 133
coevolution 54
coexposure, long-term 86
collection techniques 105
color 122
combustion
chamber 112
products 28
common response order 80
community 54
measures 60,64
natural 7,15
organization 60
competitive release 59,60
competitors 54,61
complexity
abiotic 60
biotic 60
organization 45
system 56,59
component(s) 44,46,52,53,94
acid 116
autotrophic 50
base 116
behaviors 45,59
biotic 60,62
elimination 58
high-frequency 58
inorganic 116
interactions 48,56,63
isolated 46,49,55,58,62,65
measurements 47
neutral 116
organic 116
processes 45
properties 6
related 61
responses 49
subsystems 46
test system 64
toxic 117
composite samples 138,141
compounds
See chemicals
concentration(s) 7,102,109,
117,139,141,142
addition 76–87,92,93,146
additive 86,90,91
ambient 78,89,90
effective 110
exact 39

factors 12
fixed 108
high 10,63
of liquids 111
low 8,10,113
maximum safe 114
reduction in 136
of solids 111
total 87
zero 108
See also lethal concentration
conceptual barrier 46
conditions, physical, chemical and biotic 7
congeneric homotaxis 61
Congress 37
Consent Decree 123,124
constituent(s) 79,86,88
compounds 103
concentration 106
hazardous 102
independently acting 80
mixture of 82
proportion of 95
in stream 102
toxic 102,116
Consumer Product Safety Act 37
contaminants 8,29
chemical 75
environmental 20,80
metal 81
multi- 88
organic 27
contamination 12,14,83,105, 139–142
of drinking water 29
of human beings 14
oceanic 8
preventing 106
progressive 15
single 93
of waters 81
control(s) 25,39,77,109,111,149
devices 102
engineering 121
intermediate 140
mechanism 61
negative (LC_{50}) 108–111
positive 111
of releases 55
sample 133
solvent 111
strategies 24,139,143
of substances 4,7
technology 102,103,117,121, 124,142,149
test 132
theory 45
conventional clarifiers 126
cooperative grant study 122
copper 81,82,88–93,145
-cadmium mixture 88–90,95
-nickel mixture 88,92–95
-zinc mixture 88–91,95
Council on Environmental Quality (U.S.) 6
county population size 32
coupling 48,50
court
cases 13
decree 121
hearings 13
crisis of environment 7,149
criteria pollutants 122–125
critical target site 88
cyanide 124,129,131,140
cyclones 105
Cyprinodon variegatus 110

cytotoxicity 109,114,117,133, 138,139

dangerous materials 148
Daphnia magna 110,114,115, 132–142
data
  aggregate 34
  analytical 139
  bioassay 139
  county-level 34
  interpretation 111
  raw test 111
  reporting 111
  sheets (lab) 111
  summary 34
5934 Data System 128
Dayton, OH 33
death 101,133
  acute 77,79
  rates 145
decomposition 58,59,64,65
deficiency state(s) 77,80
definitive test 108
defoliation 57,146
defoliator, geometrid 57
demographic variables 32
deposition 75
detection limit(s) 135,143
deterministic approach 20
dibromochloromethane 30
di-*n*-butyl phthalate 129,130, 139
dichloroaniline 131
1,2-dichlorobenzene 139
1,4-dichlorobenzene 130
2,2-dichlorobenzene 130
dichlorodiphenyltrichloroethane (DDT) 11–13,28,43,57,145
1,1-dichloroethane 130
1,2-dichloroethane 30
trans-1,2-dichloroethylene 129, 130
1,2-dichloropropane 130
cis-1,3-dichloropropene 130
trans-1,3-dichloropropene 130
dieldrin 28
diethyl phthalate 129,130
differential solubility 12
dilution 10
2,4-dimethylphenol 130
dimethyl phthalate 129,130
dimethyl sulfoxide (DMSO) 106,107
2,6-dinitrotoluene 130
discharge multimedia environmental goal (DMEG) values 114,116, 117
disease
  clinical 25
  control 20
  environmental 38
  heart 20,39
  incidence 27
  infectious 19,20
  initiation of exposure 33
  malignant 20
  noninfectious 21
  rate 24
dispersion 15
disposal, ground 112
dissipative system 45
dissolved air flotation capability 122
dissolved oxygen (DO) 51
disturbance
  external 45
  low-grade 77
DNA-binding agents 94

dolomite leachate 112–117
dose
  -dependent 23
  experimental 22
  high to low 23
  lethal 108–113,133
  -lethality curves 92
  maximum 108,135
  oral 31
dose rate(s)
  effective 27
  human 44,65
dose-response
  curve 80,95
  data 76,83
  lines 82,88
  relationships 21,24,25
drinking water 20,27,31,34
  constituents 28
  effect 35
  exposure variable 33
  organic contamination 29
  quality 33
  sources 30,31
Drosophila melanogaster 111
dyads, workable 147,148
dynamic(s)
  behavior 53–57
  intrasubsystem 58
  short-term 48

early warning measures 147
eastern deciduous forest 61
eastern seaboard 14
EC_{50} 110–113,133,138,139,142
ecological effects 101,105–108, 126,132
  protocol 111
  research 147
  test 106,109–111
ecological problems 47
ecological tests 114
ecological theory 62
ecology 44
  freshwater 134
  marine 134
  principles of 147
economic consequences 81
economic power 149
economic system 14
ecosystem(s) 4,43–73,132,142–149
  analysis 44
  aquatic 81,132
  component 44,54,148
  dynamics 43,47,52,61,62
  effects 44,148
  forest 59,148
  homeostasis 61
  lake 58,148
  level, total 52,59,60
  management 46
  natural 13,46,55,63–66,146
  perspective 146–149
  responses 45,52,62
  "state of the whole" 66
  stressed 50
  studies 4
  terrestrial 56,132
  theory 43–73
  urban 148
effect(s) 13,56–58,62–64,79
  biochemical 77
  biological 88,126
  biotic 7,8,11
  carcinogenic 21
  chemical 126
  direct 56
  ecological 101,102,106–108, 109–111,126,132,147

environmental 24
first-order 56
health 29,36,101,102,105,114
health, test 105–108
higher-order 56–58,62,64
indirect 56
interactive 24
lethal 86,93
physical 126
physiological 77
potential 63
somatic 10
sublethal 92,93
synergistic 4,24,132
terrestrial 108
toxic 57,63
effective concentration 133
effluent(s) 121–144
complex chemical 145
components 132
emissions 121,122
exposure 133,135
samples 130,132,135,136
secondary 127–143
specific 96
standards 121,123
tertiary 139–141
toxic 121,143
egg production, inhibition of 92
Eh 51
element(s) 50–52,63–64,88,131
biologically available 81
conservation 52
critical 51
cycles 50
loss 52
recycling 52
storage 64
embryo, developing 83
emission(s) 121,122
air 112
controls 81
effects 4
overt 76
rates 102
samples 132
standards 101
streams 102,121
endpoint 108–111
response 75–76
endrin 28
energy 146,149
-absorbing process 50
chemical bond 50
conversion processes 53,101
dissipation 50
fixation-dissipation 50
-releasing process 50
solar radiation 50
technologies 145
Engineering Science, Inc. 122
environment 8,16,25,37,75,101
ambient 21
aquatic 81
chemical 21
crisis of 7,149
disposal to 114
general 14
human 22,25
physical 55
terrestrial 51
environmental agents 77,147
environmental assessment methodology 102,104,125, 126
environmental carcinogens 19–42
environmental conditions 22,25

environmental contaminants 20,80
Environmental Defense Fund 31
environmental degradation 58
environmental disease 38
environmental effects 24
environmental exposures 27
environmental factors 6,19,20, 23,37,89,95
environmental hazards 39
environmental impact 112
environmental legislation 37, 101,149
environmental network 20
environmental parameter 91
Environmental Protection Agency, U.S. (EPA) 6,29, 37,102–104,121–125,132, 138–139
  Administrator 13
  Bioassay Subcommittee 135, 136
  Effluent Guidelines Division 121
  environmental assessment program 112
  Good Laboratory Practice (GLP) 111
  Industrial Environmental Research Laboratory (IERL) 102,122
  Process Measurements Branch 125
  Project Officer 111
  Research Triangle Park 102
  specifications 127
  technical advisors 133
  textile program 124
environmental quality 76
environmental samples 107,135
environmental systems 43
environmental tolerances 61
environmental toxicologists 77
environmental variables 89
environmental vectors 76
enzyme systems, microsomal 94
epidemiologists 38
epidemiology 19,24
  approach 24
  data 25
  investigations 25,28,31
  methods 38
  model(s) 20,24,33,34
  observation 36
  reasoning 34
  studies 4,21,23,27,44,65,148
  testing 24
equivalent response 92
Escherichia coli 138
esters 131
esthetic values 61
estuarine areas 14
ethylbenzene 129,130
bis(2-ethylhexyl)phthalate 129,130,139
etiologic hypotheses 24
Europe 16,28
Euryale 1
evacuation 8
evaluation criteria 113,123
Evansville, IN 29
evolutionary collapse 53
evolutionary histories 62
evolutionary legacy 10
evolutionary processes 7
evolutionary testing 8
evolutionary time 53
excretion 75
excretory rates 22,58

experimental cultures 109
exposure(s) 19–39,76–78,101, 121–144
  acute 76,108,111
  ambient 33
  to carcinogens 20
  chronic 20,76
  data 25,38
  duration 95
  earliest 26
  effects 77
  environmental 27
  evaluations of 19
  greater 79
  high occupational 21
  initiation of, to disease 33
  intensity of 27
  intermittent 76
  limiting 37
  long-term 19,27,77
  low-level 19,20,27,77
  mean lethal 8
  models 27,34
  multiple 75
  to organic chemicals 27
  period 83
  potentially dangerous 28
  relevant 25
  short-term 108,114
  single 24
  status 38
  variables 34
extrapolation 4,22,66,109

fall cankerworm 57
fallout 8
fatty acid groups 131
fecundity 54,57,64
Federal Water Pollution Control Act (FWPCA) 37,123
feeding activities 56
ferric chloride 122
fertilization, short-term 57
fertilized egg 83
fish 3,8,9,12,44,58,86,89,90,94, 115,132,145
  biomass 58
  mortality 58,147
  populations 13
  toxicity test 108
fission bombs 8
5934 Data System 128
flocculation/sedimentation 137, 140–142
flowrates 102
flow-through testing 63,117
flue gas 112
fluidized bed combustion (FBC) 112–117
  effluents 149
  reject 116
fluorene 130
food
  chains 12,57
  limitations 53,54
  web 15,57,101
forest
  biomass, total 60
  boreal 11
  eastern deciduous 61
  ecosystem 59,148
  landscapes 51
  productivity regulation 57
  stands 57
  succession 60
fractionation 116
Franklin County, OH 34
frequency-dependent organization 49

freshwater ecology series 135, 136
freshwater species 116
fulvic acid 29
functional group 65
functional processes 50
functional theory 46
furniture industry 61

gas 112
  flue 112
  noxious 147
  sample(s) 107,110,114–117
genes, mutant 10
genetic factors 19
genetic improvement 10
genetic predispositions 38
geochemical matrix 51
  environment 49
  inert 46
  inorganic 50
geographic differences 23
geologic forces, natural 81
germ cells 10
germination percent 111
gill tissue 88,89
glacial ice 9
gnotobiotic systems 62
Gorgons 1,2
government, federal 7,13,14
"grab" samples 103–105,112
granular activated carbon 137, 141
  adsorption 140,143
  column(s) 122,123,137
grassland productivity 59
grazing rates, regulation of 56
groundwater 6,30–33
growth 149
  inhibition 139
  stimulation 101
growth rate(s) 54
  changes 110
  increased 110
  reduction 79,93
gypsy moth 11

habitats
  aquatic 106,109
  freshwater 110,114
  marine 106,110,114
  terrestrial 106,109
haloform reaction 29,33
halogenated compounds 30
hamsters 25
hardness 90
hazardous compounds 103
hazardous constituent 102
hazardous streams 116
hazardous substances 145
hazards 19,38,80
  biotic 10
  environmental 39
  human 22
  potential 103,108
  relative 102,103
health
  care 146
  effects 29,36,101,105,114
  effects test 105–108
  of employees 38
  goals 37
  monitor 38
  of organisms 76
  protection 37
  See also human health
heart disease 20,39
heavy metals 3,52,81–99,145
  mixtures 93,95
  single 91

helium 128
hemispheres, Northern and
 Southern 9
hepatomas 31,146
heptachlor 28,130
herbivore(s) 51,56
Hewlett-Packard 5981 gas
 chromatograph/mass
 spectrometer (GC/MS)
 system 128
hexachlorobenzene 130
hexachloroethane 28
hexane 128
hierarchical cycle 4
hierarchical systems 47–49,58
hierarchy
 biological 53
 theory 46–48,55,146
hogs, keeping 5
holistic behaviors 44
holistic ecosystem responses 52
homeostatic response 45
homolog 62
Homo sapiens 61
homotaxis, congeneric 61
host characteristics 38
human(s) 5,44,63,101,109,
 146–148
 activities 15
 carcinogens 21–23,147,148
 effects research 147
 health effects 19,20,28,34,37,
 39,76
 health threat 6,13
 illness 147
 intervention 43
 life, prolonging 146
 milk 14
 morbidity 148
 mortality 148
 observational analysis 20
 observers 57
 perceptual abilities 49
 perspective 146–148
 populations 4,22–25
 safety 10
 society 61
 susceptibility 38
humic acid 29
hydraulic retention time 129
hydrocarbons 93
 aliphatic 29,117,131
 aromatic 29
 chlorinated 14,43,57
 polycyclic 28
bis(hydroxy-*t*-butyl phenol)
 propane 131

immunologic sensitivities 38
impoverishment 13
 of nature 15
 progressive 16
incipient lethal level (ILL) 86
increments of change 13
indeno(1,2,3-c,d)pyrene 130
index compound 33
indicator(s)
 cell lines 133
 organisms 132
 sensitive 58,64
 species 133
induction 94
industrial
 activities 6
 -chemical environment 19,20
 organics 94
 outfall 29
 point sources 123
 pollution 28
 records 38

wastes 7,14,16
wastewaters 124
infectious diseases 20
control of 19
infraadditive toxicity 79,82,86, 87
infrared analysis 131
ingestion 101
inhalation 36,101
inorganic(s) 94,116
compounds, low-energy 49, 50
toxic pollutants 139,142–143
insecticidal properties 11
insect toxicity test 107–113
insensitivity of test 116
intact system 58
intake water 137,138
integrated measure 51,52
integrated system 44,46,51
integrative approach 47
integrative properties 57,66
interaction(s) 80,86
antagonistic 103,116
biotic 54,55
component 48,56,63
constraining 55
critical 55
kinetic 88
metabolic 94
network 57,58,61
physiological 75
structure 53
synergistic 79,103
time-related 84
interactive effects 24
intermediate control 140
intermediate products 65
intermediate removal ability 140,141
interspecies variation 24
intervention(s) 27,145
specific 148
strategies 36
Inventory Candidate List 2
invertebrate(s) 106–117,132
aquatic 117
toxicity test 108
iodine-131 10
iron 117
level 142
residual 141
irreversible changes 101
isokinetic conditions 105
isothermal conditions 105

Japan 8,9

Kanawha River, WV 28
Kepone 3,94
ketones 29
Kiehl, S.J. 146
kinetic processes 75
Kuderna-Danish evaporator 128

laboratory
animals 21,22
experimental techniques 36
results 66
testing 24
lakes 58,146
landscape units 49
latency 25,33
latent period 22,25–27
latitudinal exchanges 9
law(s)
amendments to 149
implementing 102
new 149

of nuisance 5,16
leachate(s) 52,66,106,111–117
dolomite 112–117
leaching 47,63,106
lead 39
in paints 3
leaf litter 56
legislation 37,101,143,146,149
lethal agent 92
lethal concentration(s) (LC_{50}) 90,133,138,139,142
algal assay 146
LC/EC data 112
lethal dose (LD_{50}) 108–113,133
lethal effects 86,93
lethality
See lethal dose
lethal levels 12
lethal potency 87,95
lethal responses 95
lethal situation 88
lethal studies 92
lethal thresholds 91,93
lethal time 85,86
lethal toxicity 86,92
level(s)
ambient 81,90
of complexity 109
component 4,59,63
lethal 12
no-effect 80
safe 80
sublethal 86
subsystem 59
system 4,63
level 1 protocol 104–116
levels 2 and 3 protocol 116–117
Libya 2
lice 11
life
expectancy 5,19
span 22,25
life cycle(s) 95
stages 83
testing 106,117
ligands 81
limestone 112
limit of detection 135,143
liquid
chromatography column 131
samples 107,109
liver necrosis 31
London 36
Long Island, NY 13
longitudinal transport 9
Love Canal 6,15
low-resolution mass spectrophotometric analysis 131
lung disease 39
See also cancer

macroscopic property 59
macroscopic system 60
magnesium ions 90
mammal(s) 86,94,132
mammalian studies 109
mammalian tissues 109
management 45,55
of chemistry 15
ecosystem 46
of materials 6
model 11
purposes 148
strategies 4
system 14
of toxic substances 7,46
man-and-native view 146
man-in-nature view 147
marine ecology series 135

marine species 116
marshes 11
mathematical models 21
matrix 44,46
maximum exposure level 39
maximum permissible
  concentration 93
measures, cost effective 103
mechanisms 10,24,36,88
media, contaminated 93
medical research 19
Medusa 1–4,76,145
membranes, boundary 83
mercury 43,57,81–87,117,139
metabolic activity 133
metabolic network 47–50
metabolic pathways 110
metabolic processes 133
metabolic rate 83
metabolic system 52
metabolic transformation 75
metal(s) 76–99,124
  constituent 79
  cumulative 86
  essential 77,80
  individual 81
  mixtures 81,82,96
  mobilization 81
  nonessential 77
  toxic 140–141
  See also heavy metals
metatheories 47
methodologic problems 20
methylene chloride 106,128,
  131,139
methyl ketone 29
methylmercury pollution 147
Miami, FL 33
microbial growth 56
microbial methylation 57
microbial mutagenicity 132
microbial remineralization 58
microcosm(s) 60,146,148
  aerobic 51
  aquatic 51,60,146
  experimentation 64
  gnotobiotic 54
  laboratory 66
  terrestrial 52
microorganisms 106,132
"middle-number systems" 45,47
Midwest 30
milk 12,14
Minamata disease 147
minnow
  fathead 110,133–135,139
  sheepshead 110,133–135
mirex 94
Mississippi River 28,31
Missouri River 28
mixed deciduous forest, cycling
  in 146
mixed-function oxidase (MFO)
  systems 94
mixture(s) 82
  ad hoc 76
  binary 82
  complex 145
  inorganic-organic 96
  lethal 83
  organic 96
modeling, free body 55
models 20,24,25,33,34,39
molecule, complex 65
monitoring points 64
Monsanto Research Corporation
  (MRC) 124–127,133,137,
  138
  /EPA study 124,125,137,143
morphologies 53

mortality 35,54,84,148
patterns 23
probit 86
rates 6,20,26,27,31,32
mosquitoes, salt-marsh 11
mountain lakes 9
mouse 22,23,31,63,109
dancing 147
Muller 11
multimedia filter 122,123,137
multimedia filtration 122-126, 137–143,146
multiple exposure 75
multiple regression analyses 31
multiple toxicity 75–99,148
nonadditive forms 80,81
mutagen(s), chemical 135
mutagenicity 106,114,117,132, 135,138,139
calculation 112
microbial 132
potential 132
mutations 10,108
mycorrhizal components 54
Mysidopsis bahia 110

naphthalene 129,130
National Academy of Sciences 30
National Cancer Institute 31
National Organics Reconnaissance Survey 30, 32,33
National Research Council, Water Subcommittee 96
natural communities 7,15
natural constraints 55
natural context 54
natural ecosystem 13,46,55,63–66,146
natural environments 50,51,62, 65,66
natural organic matter 29
Natural Resources Defense Council et al. v. Train 123
natural systems 49
negative health effect 36
negligence 43
neoplasms 32,36
See also cancer
nerve cell 54
net productivity, autotrophic 56
neurological disorders 39
New England 9
New Orleans, LA 31
New York Times 13
Niagara Falls, NY 6
nickel 82,86–88,92,93,117,148
-vanadium mixture 82,86,87, 95
nitrogen 15,50
mineralization 64
recycling 59
2-nitrophenol 130
4-nitrophenol 130
n-nitrosodiphenylamine 130
n-nitroso-di-*n*-propylamine 130
noncumulative agent 86
noninfectious diseases 21
nonlethal responses 109
nonlinear relationships 25
nonspecific action 83
nonvolatile organics 124
North America 9
eastern 11
Northern Textiles Association 122
nuclear war 5
nuisance 5,16

nutrient(s) 15
  assimilation 56
  available 59
  cycling 56,59
  exchange 146
  freed 59
  limiting 110
  loss 56
  release 148
nutritional deficiencies 38

Occupational Safety and Health Act 37
ocean(s) 14
oceanographic surveys 9
octopus 147
odor problems 28
Ohio 31,32
ontology 94
operating conditions, effects of 103
operating cycle 104
organ
  activity 77
  targets 83
organic(s) 94,105,116
  acid extractable 128
  base/neutral extractable 128
  chemical constituents 20
  compounds 29,49,128
  contaminants 28
  extraction of 106
  -free water 128
  industrial 94
  -inorganic mixtures 95
  molecules 81
  nonvolatile 124
  pollutants, toxic 138–141
  precursor 29,33
  species 136
  structures 50
  synthetic 3,26,27
  toxic pollutants 142
  volatile 124,126,128
  See also natural organic matter
organism(s) 7,54,65,93,101, 106,109,147
  aquatic 88,94,132
organization
  biological 66
  community 60
  complexity 45
  frequency-dependent 49
  levels of 53
  rules 55
  states 51
  structure 56
organohalogen by-products 29
organophosphates 94
osmotic imbalances 116
outcome 25,27
output
  aggregate 48
  rate 59
ozonation 137,140,141,146
ozonator 123,137

Pacific, western 8
paints, lead-containing 3
paradigms 75–96
paraffin/olefinic 131
parallelism 82,95
particles
  airborne 9
  deposition of 9
  suspended 105–107,112
  transport of 9
particulate(s) 105
  coarse 112
  fine 112

respirable 109
samples 114–117
solids 112
pathogenic mechanisms 21,25
pathological effects 77
pathways, biogeochemical 50, 148
pE 51
penguins 11
pentachlorophenol 130
peristaltic pump 128
permissible range 76
Perseus 1,2
persistence 10–13
persistent organic structures 46
persistent toxins 10
perturbations 51,81
pesticides 7–16,124,128
concentrated 57
control of 7
mixtures 94
organic 93
organochloride 28,94
production 3
short-lived 16
pH 29,33,50,51,90,122,128
-pE phase space, two-dimensional 63
phagocytic cells 109
pharmaceuticals 3
pharmacokinetics 146
pharmacological toxicity observations 109
pharmacology 22
phase space, pH-DO 51,60,146
phased approach 103,112
Phaseolus vulgaris 110
phenol 122,124,129,130,139
alkyl 131
phosphorus 15,51,58
photoperiod, alteration of 51
photosynthesis 64
photosynthetic pigments 49
phthalate esters 131
physical effects 126
physical environment 55
physical studies 102
physicochemical processes 57
physiological effects 77
physiological interactions 93, 94,96
physiological-level data 55
physiological mechanisms 44
physiologies 110
phytoplankton populations 59, 60
pilot
locations 137–138
studies 104,110,117
pilot plant
effluents 139
field testing 123
mobile 122,136,137
program 125
Pimephales promelas 110
piperonyl butoxide 94
plant(s) 54,106,109,111,132
relationships 147
stress ethylene test 104,107–110,113–115
point sources 76,123
poisoning 3,147
poisons 80,82
polarity 131
policy
determinations 27
for intervening 21
-makers 38
management 13
political compromise 7
political power 149

pollen grains 9
pollutant(s) 101
chemicals 75,94
coexisting 76
concentrations of 102
criteria 122–125
individual 82
known 103
specific 103,136
toxic 138–143
pollution 7,28
air 6
control technology 102
multimetal 93
problems 5
of waters 94
polychlorinated biphenyls (PCB) 9,14,124
polyelectrolytes 122
polyethylene containers 105
polymers, cationic 143
population(s) 53,65
approach 66
aquatic 81
biological 57
consumer 56
elimination of 59
faunal 58
heterogeneous 25
high-risk 38
-level data 55
-level measure 64
measures 60
microbial 60
studies 24
surveillance 38
potassium 117
potatoes 12,13
potency 77,79,87,95
potential threat 94
potential toxic effects 65
potentially hazardous effects 3
potentially toxic chemicals 2
potentiation 80,94
precipitation 9
precoagulation 140,141,143
predator(s) 54
-prey 56
predictability, judgment of 22
presence-absence records 60
pressure 102
streams 104,105
preventive medicine 19
primary prevention 37
primary productivity rates 56
principal-components ordination 60
priority pollutant(s) 123–130
analysis 124,125
concentrations 124,125
metals 129,131
organic 128,130,136
probability 20,34,39,53
low 22
process
anabolic 45,50
biochemical 51
catabolic 45,50
engineering 102
streams 145
variations 117
productivity
net, autotrophic 56
total 58–60
products, intermediate 65
profits 149
protein(s)
denaturation 83
enzymatic 81
structural 81

protocols 101,116,127,138
algae test 110,133
of biological tests 4
terrestrial 110
toxicological testing 44
public health
hazard 28
programs 11
pure solutions 88
pyrene 130

quality control and quality assurance (QC/QA) 111
quality, environmental 76
Quebec Department of Education 96

rabbit(s) 54
alveolar macrophage (RAM) test 107–109,113–115
radiation 8
ionizing 10,11
radioactivity 8–10,15
radionuclides 9,12
rains, heavy 6
random connections 53
random interactions 53
rangefinding test 108
ranking
relative 113,135
of streams 103
rat(s) 22,25,31,133,145
biological tests 135
rate
changes 58
constants 47
-dependent processes 64
processes 50,58,64,65
raw water 28–30,127–131
reactor/clarifier 122,123,137
recycling
nitrogen 57
nutrient 57
reduced growth 108
Registry of Toxic Effects of Chemical Substances 2
regression 32,34,146
regulation 39,43,149
density-dependent 53
toxicity-based 81
regulatory actions 63
regulatory agencies 20,39
regulatory concern 27
regulatory purposes 44
regulatory statutes 36
reindustrialization 16
relative abundance 59,64
relative potency 82,92
relative ranking 135
relative response 135
release limits, acceptable 102
remineralization 51
removal ability, intermediate 140,141
renal epithelial tumors 31
reproduction response 93
reproductive capacity 145
reproductive failures 12
research design 24
Research Triangle Park, NC 102,122,126
residual coagulant concentrations 141
residual inorganic coagulant 143
residual polymers 142
residues 11–14
resistant strains 12
resource(s)
exploited 8

shared 14
Resource Conservation and Recovery Act (RCRA) 101
respiration 83
response
addition 146
adverse 77
biogeochemical 52
carcinogenic 44
component 49
curve 82
ecosystem 45,52,62
homeostatic 45
integrated system 44
joint 79
lethal 95
long-term 44
nonlethal 109
observed 82,84
pattern 81
predicted 84
sublethal 95
system 49
time-related 86
toxic 44,65
whole-organism 79
restitution, limited 6
retention time, hydraulic 129
risk(s) 19–24,27,31
additive 20
assessment 23
carcinogenesis 146
estimates 22,34
of exposure 24
high 76
high-, groups 24
high-, individuals 38
human 20–22,27–28
measurable 28
synergistic 20
rodent(s) 44
acute toxicity test 106–109, 113–115
Rongelap Atoll 8
root elongation test 107,108,110
root length 111
route of uptake 95
Ruckelshaus 13

Safe Drinking Water Act (SDWA) 27,29,37
Safe Drinking Water Committee 30
safe level of usage 13
"safe" range 76
safety factor 83
salicyclamide 23
Salmo gairdneri 86
Salmonella 108
Salmonella typhimurium 138
salt concentration, high 116–117
sample(s) 104–106,113–116
analysis techniques 139
collection 104
containment 113
environmental 107,135
gaseous 105,106,116
integrity 105
liquid 105,106,113
preparation 104–105
second cyclone 116
solid 106,113
storage 105
transport 104–105
sampler, automatic 126–129
sampling and analytical procedures, manual of 124
sampling
line (Teflon) 105

probe 104
procedures 103
scientific evaluation 39
scientific evidence 21
scientific input 39
scientific investigations 39
scientific testing 39
screening study 126,135
seals, Antarctic crabeater 11
second cyclone
catch 114,115,117
filter catch 112
leachate 115
secondary effluent(s) 127–143
sample 127
solution 133,134
toxicity 125,135–137
treatment 140
secondary treatment plant 125, 127,131,137
sedimentation 137,140–142
seed germination–root elongation test 108,110,113
Selenastrum capricornutum 110
selenium 87,117,139,148
sensitive species 44,61,135,142, 147
sensitization 80
sequential processing 65
sewage 7
collection system 16
shrimp, grass 132–135
similar action, hypothesis of 82
single agents 79
single cell line 138
single point in time 104
single population 61,64
single-species systems 44
single-species test systems 62
single substance toxicity 76
Skeletonema costatum 110
skin, of mouse 63
slip-stream 137
smallpox 19
smoke 28
smokers 38
smoking, cigarette 25
snakes 2,146
Snow (of London) 36
Snyder column 128
social goals 37
sodium ions 90
sodium sulfate column 128
soil(s) 13,59
cores 52,60
microcosm 133,134
organisms 133
relationships 147
system 60
solar radiation energy 49
solid(s) 105,107
ash materials 112
residues 112
samples 109
streams 114
wastes 111
solubility, differential 12
solutions, discrete 80
solvent
exchange 106,107
samples 105
sorbent
column 105–106
material 112
source assessment sampling system (SASS) 102–105, 112
spatiotemporal scale 49
species 50–61

algal 110
changes 61
composition 59
eliminated 53,61
endangered 61
interactions 64
introduction 54
invertebrate 110
populations 50
replacement 60,61,64
vertebrate 110
spectral analyses 52
spectral properties 60
spray program, aerial 11
spruce budworm 11
stability 53,59
stack 104
standard test protocols 111
standing crop 59
"state of the whole" 46
static acute procedures 110
statistical analysis 24,34
status
age 38
general health 38
race 38
sex 38
steady-state conditions 138
steady-state properties 48
steam power generation 112
Stheno 1
stimulation concentration (SC) 110
stirring regime 51
stock cultures 62
storage compartments 148
strategies of intervention 37
stream(s) 116
hazardous 116
priority 103
stress, response to 110
strontium-90 10
structural detail, microscopic 59
structural organization 62
study design 34
styrene 23
subfractions 116
sublethal agent 92
sublethal effects 92,93
sublethal levels 86
sublethal responses 95
sublethal thresholds 88,91,93
substrates 58,66
subsystems 46,48,59
subthreshold level 93
subthreshold presence 84
subthreshold substances 76
succession 60
sulfate 112
sulfide 122
sulfur 112
sulfur dioxide 25
suppression 94
supraadditive 79,82,86,88,91–95
surface aerators 126
surface/groundwater dichotomy 33
surface water 30–32
suspended particles 105–107, 112
swordfish meat 57
synergism/antagonism approach 143
synergistic effect(s) 4,24,132
synergistic interaction(s) 79, 103
synergistic relationships 25

synthetics 3
system
  behaviors 47,49,58–61
  biogeochemical 46,49,51
  complexity 56,59
  -component problem 46
  destabilized 54
  dynamics 63,66
  evolution 59
  high-order 53
  -level effects 64
  -level measure 52
  level, total 65
  organization 47–48
  properties 64
  response 64
  stability 53–54,59–60
  stress 64
  structure 47,49
  unstable 53
systems engineering 45

target
  organism 93
  site, critical 88
  tissue 75–78,85
taxpayer, cost to 6
techniques and tools 146,147
Tedlar sampling bags 104,105
temperature 29,102
temporal sequence 34
temporal variations 105
teratogenicity 117
terrestrial effects 108
terrestrial protocol 110
terrestrial tests 110,111
tertiary effluents 139–141
tertiary treatment
  system(s) 125–126,136–138, 141–143
  technologies 124,132,135–138
  unit operation 125
test
  algal 110,133
  aquatic 105–106,110,113–114
  bioassay 124,135,139,140
  biological 4,105–107,113,121–125,131,132,142,146
  chemical 149
  condition 132
  definitive 108
  ecological 114
  high doses 22
  insensitivity 116
  organisms 80,83,86,107–109, 117,132
  populations 82,92
  procedures 65
  sensitive 65
  species 22,108,139
  system(s), intact 62–66
  terrestrial 110,111
  toxicity, algal 108
  in vitro 106,108,109
  in vivo 22,106,108,109
  See also testing protocols
tester strains, bacterial 135
testing
  flow-through 63,117
  long-term 117
testing protocols 58,62–66,103–104
  algal 110,133
  biological 117,133,143,149
1,1,2,2,-tetrachloroethylene 130
textile(s)
  manufacturing industry 118, 122–144

manufacturing plants 122–144,145
plant wastewaters 121–128,142,149
thallium 136,139
theoretical rationale 66
thermodynamics 45,50
threshold(s) 7,13,77–80
concept 15
lethal 91,93
limit 106
sublethal 88,91,93
tolerance 76,77
toxicity 81
See also subthreshold
time
constant 48
elapsed 116
function of 26
tissue
concentration 23
culture 53,54
function 77
targets 83
titration 51
tobacco 21
tolerance 7,80–83,95
lethal 83
thresholds 76,77
toluene 129,130,139
-sulfonyl groups 131
total suspended solids (TSS) 122
total system (properties) 60–61,148
toxic action sites 78,79
toxicant(s)
application of 43
cumulative 86
disposal of 43
individual 80
potency of 79
toxication 13–16
toxic chemicals 37
toxic constituents 102,116
toxic discharges 46,65–66
toxic effects 57,63
toxic effluents 121,143
toxic inorganic pollutants 139,142–143
toxicities, joint 81
toxicity
acute 110,133,135,139–142
animal 4
aquatic 107–108
baseline 125
behavioral signs of 84
biological 114
calculation 112
cause of 116
curves 85–86
data, chronic 30
decrease in 90
evaluation of 112
heavy metal 87–90
infraadditive 79,82,86,87
lethal 86,92
lower levels of 143
to mammals 11
measurements 123
multiple 75–99,148
net 116
pattern 79,95
pattern response addition 80
potential 132,145
ranking 136
reduction in 90,125,135–136
relative 136

removal 125,131,140,143
response patterns 75
screening 136
of stream 103
study 124,132
testing 96,132
toxic metals 140–141
toxicological effects 121
toxicological impact assessments 76
toxicological problems 3
toxicological procedures 65
toxicological studies 28
toxicological testing 44
toxicologists 38,77
toxicology 44
methods 21
"new" 21
testing 37
toxic organic pollutants 138–141
toxic pollutants 138–143
toxic potential 2
toxic releases 16
toxic responses 44,65
toxic streams 126
toxic substances, defining 37
Toxic Substances Control Act (TSCA) 2,6,37
toxic unit formula 81,88,93,148
toxic unit standard 83,86,93
toxins, persistent 10
tracers 9
transfer patterns 148
transport
aerial 12
atmospheric 10
pathways 88
treatability
studies 122
tests 123
treatment
modes 123
plant, biological 137
systems 137–140
technology 121
tri-*t*-butyl benzene 131
1,2,4-trichlorobenzene 139
1,1,1-trichloroethane 130
trichloroethylene 129,130
trichlorofluoromethane 130
2,4,6-trichlorophenol 130
trihalomethane (THM) 29,32, 33
TRIS 3
trophic dynamics 146
trophic levels 53
troposphere 9,12
trout 86
tuberculosis 19
Tucson, AZ 33
tumor(s)
See cancer
tumorigenesis 94
tumorigenic outcome 22
tumorigenic responses in humans 148
turnover times 59
Tygon tubing 128

unicells 109
United States 3,6,9,11,13,14,20, 23,28
District Court of Washington, DC 123
eastern 54
Fourth Circuit Court of Appeals 122
International Trade Commission 2

uptake 51
  rates 88

vanadium 86,87
variables 38,63
  biogeochemical 64
  biotic 89
  demographic 32
  environmental 89
  independent 31
  master 50
  temporal 27
variances, normally distributed 82
variation, interspecies 24
vectors 76
vertebrates 109,110,132
viability test 133
vinyl chloride 21
vinyl stearate 131
visual damage parameters 110

warning, published 11
Washington, DC 29
waste(s) 7,14,16
  chemical, disposal 3
  industrial 101
  products of technology 8
  recovery of 16
  release of 101,109
  system 117
  technology, low-level 16
  treatment study 127
waste streams 102–117,149
  complex 108–109
  ranking of 102,112
  samples 105
  toxicity of 113
wastewater(s) 143
  industrial 124,132
  municipal/industrial 132
  samples, raw 134,137
  secondary 135
  streams 140
  toxicity 137
  treatment 136
water
  bodies 16,83
  column 58
  consumption 34
  deep well 30
  dilution 132
  effect 34
  finished 28,30
  hardness 88–91,95
  intake 127,138
  pollution 6
  samples 124
  soft 88–91
  sources 33
  system, domestic 16
  treated 28
  variable 31
  See also drinking water; groundwater; raw water
water quality
  criteria 93,96
  management 88–89
  objectives 79,91
watershed studies 51
water supplies 29
  municipal 28
  potable 27
  public 29
water treatment
  plant 34
  practices 33
  procedures 29
weeds, aquatic 88

wells 13
whole organism responses 95, 109
whole (total) system responses 51,52
wilderness regions 76
world population 5,6
World War II 11
  post- 27

zebra fish 83–85,92
zero concentration 111
Zeus 1
zinc 81–84,88,95